Sebastian Clauß

Thermal stability of
1C PUR adhesives

Sebastian Clauß

Thermal stability of 1C PUR adhesives

Structure-property relationships of one-component moisture-curing polyurethane adhesives for timber construction

Südwestdeutscher Verlag für Hochschulschriften

Impressum/Imprint (nur für Deutschland/only for Germany)
Bibliografische Information der Deutschen Nationalbibliothek: Die Deutsche Nationalbibliothek verzeichnet diese Publikation in der Deutschen Nationalbibliografie; detaillierte bibliografische Daten sind im Internet über http://dnb.d-nb.de abrufbar.
Alle in diesem Buch genannten Marken und Produktnamen unterliegen warenzeichen-, marken- oder patentrechtlichem Schutz bzw. sind Warenzeichen oder eingetragene Warenzeichen der jeweiligen Inhaber. Die Wiedergabe von Marken, Produktnamen, Gebrauchsnamen, Handelsnamen, Warenbezeichnungen u.s.w. in diesem Werk berechtigt auch ohne besondere Kennzeichnung nicht zu der Annahme, dass solche Namen im Sinne der Warenzeichen- und Markenschutzgesetzgebung als frei zu betrachten wären und daher von jedermann benutzt werden dürften.

Coverbild: www.ingimage.com

Verlag: Südwestdeutscher Verlag für Hochschulschriften GmbH & Co. KG
Heinrich-Böcking-Str. 6-8, 66121 Saarbrücken, Deutschland
Telefon +49 681 37 20 271-1, Telefax +49 681 37 20 271-0
Email: info@svh-verlag.de

Approved by: Zürich, ETH, Diss., 2011

Herstellung in Deutschland (siehe letzte Seite)
ISBN: 978-3-8381-3385-0

Imprint (only for USA, GB)
Bibliographic information published by the Deutsche Nationalbibliothek: The Deutsche Nationalbibliothek lists this publication in the Deutsche Nationalbibliografie; detailed bibliographic data are available in the Internet at http://dnb.d-nb.de.
Any brand names and product names mentioned in this book are subject to trademark, brand or patent protection and are trademarks or registered trademarks of their respective holders. The use of brand names, product names, common names, trade names, product descriptions etc. even without a particular marking in this works is in no way to be construed to mean that such names may be regarded as unrestricted in respect of trademark and brand protection legislation and could thus be used by anyone.

Cover image: www.ingimage.com

Publisher: Südwestdeutscher Verlag für Hochschulschriften GmbH & Co. KG
Heinrich-Böcking-Str. 6-8, 66121 Saarbrücken, Germany
Phone +49 681 37 20 271-1, Fax +49 681 37 20 271-0
Email: info@svh-verlag.de

Printed in the U.S.A.
Printed in the U.K. by (see last page)
ISBN: 978-3-8381-3385-0

Copyright © 2012 by the author and Südwestdeutscher Verlag für Hochschulschriften GmbH & Co. KG and licensors
All rights reserved. Saarbrücken 2012

Preface

Bonding of wood today is one of the most important joining techniques in timber construction and allows constructions to be built that would otherwise only be feasible with stronger materials such as steel or reinforced concrete. In times of limited natural resources, wood, as a renewable resource, is becoming more and more important for constructors, architects and engineers, due to its material specific advantages, such as its high strength to density ratio along with its positive eco-balance.

Since the evolutional development of wood does not fit engineering demands in all cases, solutions must be found to optimize the material for civil engineering. The invention of glulam by Otto Hetzer in 1906 was a milestone for structural wood construction. Due to bonding, it was possible to improve the properties of the material significantly by grading and removing defects to apply the strength properties to full capacity.

Apart from wood, the adhesive is the second integral component. During Hetzer's time, one relied on high class casein adhesives made from acid milk casein, which was added with calcium compounds. This biological adhesive is interesting from an ecological view, however it is not resistant against moisture and therefore not capable of resisting delamination loads. Synthetic resins, such as phenol-resorcinol-formaldehyde (PRF) and melamine-urea-formaldehyde (MUF), which furthermore can be cost-efficiently produced from petrochemical raw materials, quickly replaced biological adhesives and are currently the most used adhesives for the bonding of timber.

In the late 1980's, a new type of reactive adhesives based on polyurethane was developed and introduced on the local market. The polymerization of polyurethanes goes back to developments made by Otto Bayer in 1937. By using polyurethane as an one-component adhesive, isocyanate-terminated polyurethane-prepolymers react with water from the substrate or the surrounding air humidity. Thus the system is completely different to water-based polycondensation adhesives such as urea-, melamine- or phenol-resin in that they include water as a solvent and use an additive hardening component for curing at room temperature.

Even though one-component polyurethane (1C PUR) exhibits favorable characteristics, such as its ductile material behavior or its modifiability in terms of reactivity and viscosity, deficits existed regarding its thermal stability in comparison with PRF or MUF. In the past, the resistance of 1C PUR in a fire situation has been doubted. As a result, research facilities, standardization organizations and the industry have strived towards developing new standards and test methods. The developed tests partly require that the adhesive must perform equally or better than the bonded wood. 1C PUR adhesives, developed in the 1990's, were not able to pass these tests.

Knowledge regarding the thermal resistance of wood bonding with 1C PUR was limited and available publications on the topic were mostly based on commercially available adhesives without knowledge of the chemical composition. Due to this lack of suitable data, a research project by the adhesive producer Purbond in Switzerland and the raw material producer Bayer MaterialScience in Germany together with the Wood Physics Group of the Institute for Building Materials at ETH Zurich, was initiated to increase the state of knowledge of bonding with 1C PUR. Within this dissertation, basic investigations on the structure-property relationships of one-component moisture-curing adhesives under thermal load were initiated to create a basis for new developments by the adhesive producers.

During the development of this thesis, I received a immense professional and moral support from many people. First of all I would like to thank my supervisor Prof. Dr.-Ing. Peter Niemz for his strong commitment in this project, but also for his professional and personal support, particulary in difficult times. Moreover I would like to thank my co-examiners Prof. Dr. sc. techn. Mario Fontana (ETH Zurich, Institute of Structural Engineering, Chair of Structural Engineering - Steel, Timber and Composite Structures) and Prof. Dr. nat. techn. Rupert Wimmer (University of Natural Resources and Life Sciences Vienna, Institute of Wood Science and Technology) for their expertise.

Furthermore, I have to thank our industrial partners from Purbond, represented by Walter Stampfli and Bayer MaterialScience, represented by Dr. Heinz-Werner Lucas for the funding of this project. Additional thanks go to the Fund for the Promotion of Forest and Wood Research for their financial support. Moreover, I appreciate the professional support in many experiments and analyses by Purbond, Bayer MaterialScience and CURRENTA, represented by Dr. Alexander Karbach. Very special thanks go to the initiators of this project, Dr. Eduard Mayer, Dr. Walter Meckel, Dr. Joseph Gabriel and Prof. Dr.-Ing. Peter Niemz who set the ball rolling in 2005.

Further thanks go to my coauthors Karin Allenspach, Dr. Dirk J. Dijkstra, Dr. Joseph Gabriel, Dr. Matúš Joščák, Dr. Alexander Karbach, Oliver Kläusler, Dr. Mathias Matner, Dr. Walter Meckel and Prof. Dr.-Ing. Peter Niemz for their excellent collaboration on the published scientific papers and conference contributions.

Many warm thanks also to my colleagues at ETH and our partner organizations for the pleasant cooperation, especially to Philipp Haß and Dr. Daniel Keunecke for the numerous scientific discussions and their critical proof-reading of the papers, Dr. Mathias Matner, Dr. Christian Wamprecht and Dr. Carlos Amen for the prepolymer synthesis and adhesive formulation, Dr. Dirk J. Dijkstra for the great support during the DMA measurements, Gabriele Peschke for her support at the ESEM, Stéphane Croptier for the preparation of microtomed specimens, Edmund Risi for his support during the sample preparation, Gabriele Linden for her support during the AFM measurements, Thomas Schnider for his technical support and Sonia Paget and the members of the Bayer language service for their English corrections.

A very special thank you to my parents for their immense and continuous support and to my partner Denise for her manifold support and encouragement during the last years of the doctorate and finally to our son Oskar for the joy he brings to us each day.

Sebastian Clauß

Dedicated to my father Christian Clauß
(1952-2010)

who inspired my interest in wood

Contents

Summary	**9**
1 General Introduction	**11**
1.1 Problematic and research gap	11
1.2 Motivation	15
1.3 Main research objectives and arrangement of this thesis	16
1.4 Specific research objectives addressed in the papers	17
2 Background to the adhesion of wood	**19**
2.1 Wood as an adherend	19
2.2 Adhesion of wood	20
2.3 Wood properties affecting adhesion	23
2.4 Penetration of adhesives – the interphase region	26
2.5 Chemistry of 1C PUR adhesives and specific properties affecting adhesion	27
3 Main investigations	**31**
3.1 Paper I	33
3.2 Paper II	49
3.3 Paper III	73
3.4 Paper IV	91
3.5 Paper V	97
4 Synthesis	**113**
4.1 Main findings	113
4.2 Assessment of the significance of this study	114
4.3 Potential for future research	115
4.4 Potential for economical issues	116
4.5 Outlook: Some general statements	117
References	**119**

Summary

The objective of this thesis is to investigate the structure-property relationships of one-component moisture-curing polyurethane (1C PUR) adhesives under thermal load. Therefore, the chemical composition of PUR prepolymers was systematically varied. Along with the cross-link density, the content of urea and urethane hard segments was specifically changed. Furthermore, the functionality of the prepolymers was adjusted by either the isocyanate (NCO) or the polyether component. Different epoxides were used for the synthesis of the polyethers. The cured prepolymers were tested as both free films and in bonding, in combination with different wood species as a function of the temperature.

In a second step, prepolymers with prospective properties were selected and formulated to adhesives with an open time of about 90 minutes. The properties of these adhesives were compared with the underlying prepolymers by the use of several investigation methods. Thus, not only macroscopic tests were performed, but also micro-mechanical tests such as nanoindentation, spectroscopic analysis such as fourier transform infrared spectroscopy (FTIR), and microscopic investigations on the micro and nano scale by means of environmental scanning electron microscopy (ESEM) and atomic force microscopy (AFM).

To reach the goal of increased thermal stability, different types of organic and inorganic filler materials with the most promising set of properties were incorporated into the prepolymer. Prepolymer integrated fillers (polyurea dispersion and styrene-acrylonitrile) and adhesive dispersed filler materials, added after the prepolymerization process (chalk and polyamide) were used. The mechanical properties of the filled systems were then tested by means of films and bonds with beech according to DIN EN 302-1 (2004) in the temperature range from 20 to 200°C. Furthermore, microscopic investigations in combination with energy-dispersive X-ray spectroscopy (EDX) were carried out to investigate the penetration behavior of the adhesives into the wooden substrate.

The results clearly showed that the structural composition of the prepolymers significantly influence the thermal stability of the bonds. In particular the urea hard segments always have a positive effect on thermal stability, whereas the urethane hard segments have a positive effect only at standard climatic conditions. The cross-link density plays a minor role in the normal temperature oscillation range, however, at high temperatures such as during a fire, cross-link density is of significant importance. The results of the prepolymers with varying functionality showed that the variations by the polyether or NCO components had no effect. The use of ethylene oxide ethers turned out to be disadvantageous in regard to high temperatures.

Summary

The formulation showed no influence on the mechanical properties of the adhesives, neither in the macroscopic nor in the microscopic range. The increase of viscosity and reactivity, however, caused significant differences in the penetration behavior of the adhesive within the substrate. Thus, clearly better bonding results could be obtained with the formulated systems. In comparison with amino- and phenoplastic adhesives, the 1C PUR adhesives displayed significantly lower values in hardness, strength and stiffness. The stored energy, however, was clearly higher for 1C PUR due to its ductile material behavior.

Through the addition of filler materials, the thermal stability of the adhesives could be significantly increased. Therefore, the less cross-linked systems with low stiffness values offered particularly favorable potential. Between the organic fillers, only small differences in the efficacy were obtained, especially at high temperatures. However, the polyamide used displayed a higher tensile shear strength at standard climatic conditions, but with reduced cohesive strength and stiffness. The best results could be obtained with a filler material content of about 7.5 %.

In summary, significant findings to the structure-property relationships of 1C PUR adhesives could be determined, which have been directly integrated in the development of new systems by the industrial partners. As a result, marketable adhesives could be produced that comply with the high demands of ASTM D 7247 (2007) regarding thermal stability. Besides the economical potential, the results can also be used for the development of failure behavior simulations of wood bonds during fire scenarios. The general statements regarding the thermal stability of 1C PUR adhesives could be adjusted by systematic investigations of their influence on structural parameters. Furthermore, starting points for continuative investigations could be identified, specifically in the area of filler materials and also adhesion between 1C PUR and wood. Through the work presented, it could be shown that by optimized combinations of prepolymer and formulation in addition to filler materials, bonds can be realized that are fairly comparable to PRF adhesives.

1 General Introduction

1.1 Problematic and research gap

The use of adhesives in timber engineering is highly regulated worldwide, since application of unsuitable systems creates a high potential for danger. Past tragedies have clearly shown that failure in bonding can cause catastrophic results for human beings. The collapse of the timber roof construction of an ice-rink in Bad Reichenhall, Germany, claimed the lives of 15 people. The collapse occurred as a consequence of a number of factors, however, the crucial factor was the wrong type of adhesive (Bautechnik, 2006). Due to different types of loading (Fig. 1.1.1), all of which collectively impact on a structural element, adhesives must meet the highest requirements (Table 1.1.1) in order to sustain the structure and ensure its integrity.

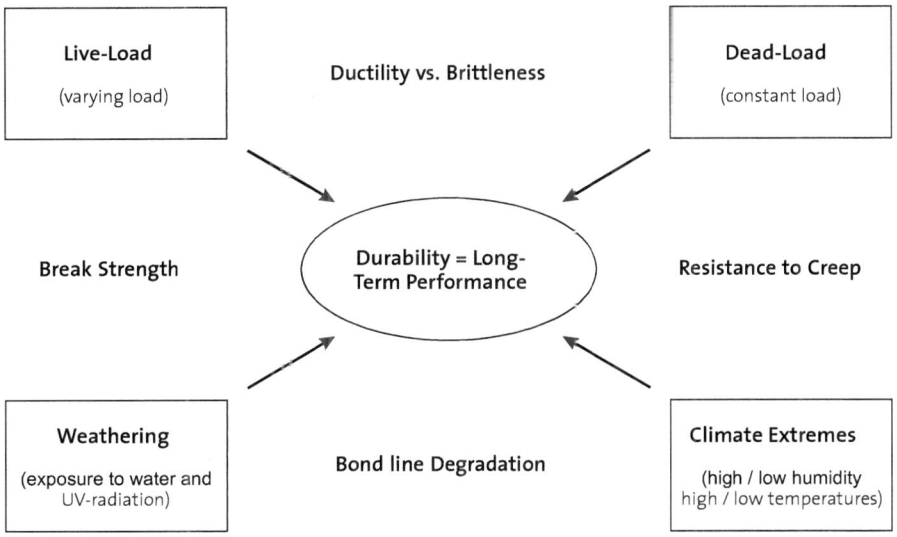

Fig. 1.1.1 Loads affecting structural adhesives (Gerber, 2008)

However, in the adhesive sector there are no consistent, but rather only local (related to economic areas or countries) standards (e.g. SNV, EN, ASTM, CSA, JAS, etc.), which have to be met to account for the adhesive usage. In Europe, the design of timber buildings and civil engineering works are regulated by the Eurocode 5 (DIN EN 1995-1-1, 2010), which uses the limit-state concept (König, 2005). The specific requirements for glued laminated timber and glued solid timber are furthermore constituted in (DIN EN 14080, 2011). The adhesives used for load bearing structures must allow the assembly of durable joints in glulam products for the required service class according to DIN EN 1995-1-1 (2010).

The classification and performance requirements for adhesives used for load bearing timber structures are classed according to the adhesive types; (1) phenolic and amino plastic adhesives (DIN EN 301, 2006), (2) one-component polyurethane (1C PUR) adhesives (DIN EN 15425, 2008) and emulsion-polymer-isocyanates (EPI). The separation was performed in 2008 after it was accepted that, based on the different chemistry of 1C PUR adhesives, their performance is fundamentally different to phenolic and amino plastic adhesives and therefore the standards had to be adapted to this type of adhesive.

The adhesive list (MPA-Liste, 2011) of the Otto-Graf-Institute in Stuttgart, Germany provides information about currently accepted adhesives according to (DIN 1052, 2008) (former document of (DIN EN 1995-1-1, 2010)) and general construction regulation approvals. Among them are resorcinol-formaldehyde (RF), phenol-resorcinol-formaldehyde (PRF), EPI, melamine-formaldehyde (MF), one- and two-component polyurethane (PUR), epoxy (EP) and urea-formaldehyde adhesives (UF). The criteria for acceptance of an adhesive for use in timber engineering are mostly based on testing combinations of different loads that occur in practical applications:

Dead load is an intrinsic invariable load on a structure, such as the bridge's own weight. Permanent loads attached to the structure may also be included. In particular, time dependent behavior of the structure is of special importance.

Live-load is characterized as a moving variable load that a structure carries in addition to its dead load, such as moving traffic on bridges or wind load on roof structures.

Weathering describes the exposure of a structure to precipitation, UV and infrared radiation, and exhaust gases. These influences cause chemical degradation of wood and adhesives similarly.

Climate extremes, such as temperatures and humidities beyond the normal range, impact structures in terms of chemical and physical bond line degradation and also in terms of internal stresses as a consequence of swelling and shrinking.

Interactions of effects, especially in the case of cycles of mechanical loads and moisture, lead to mechano-sorptive creep that causes higher deformations compared to the superposition of single influences. Furthermore, chemical aging effects can be accelerated under mechanical loads.

Looking at European standards requirements, it is obvious that tests for adhesives are predominately oriented to test the resistance against mechanical loads and moisture extremes. High temperatures or

Table 1.1.1 Load categories and standards for testing

Live-load	Determination of bond strength at standard climatic conditions	DIN EN 302-1 (2004)
Live-load and climate extremes	Determination of bond strength at high moisture	DIN EN 302-1 (2004)
	Determination of bond strength at elevated temperatures	DIN EN 15416-2 (2008)
	Determination of the effects of wood shrinkage on the bond strength	DIN EN 302-4 (2004)
Dead-load and weathering	Determination of bond strength after temperature and humidity cycling	DIN EN 302-3 (2006)
Dead-load and climate extremes	Determination of fatigue strength under static load at cyclic climate conditions	DIN EN 15416-2 (2008)
	Determination of creep deformation at cyclic climate conditions	DIN EN 15416-3 (2010)
Climate extremes	Determination of resistance to delamination	DIN EN 302-2 (2004)

fire scenarios are not included. The highest temperature in the tests is 100°C, applied as boiling water over a 6 h duration (DIN EN 302-1, 2004). In dry state, the highest temperature is 70°C, maintained for two weeks under constant load (DIN EN 15416-2, 2008). Currently the European standards for adhesives used in load bearing constructions hardly provide information about their performance at elevated temperatures, nor do they give a classification regarding their suitability in structural fire design. The fire classification of construction products and building elements is determined by full size fire resistance tests of load-bearing elements according to DIN EN 1365-3 (2000) and by calculations according to DIN EN 1995-1-2 (2010), where either the reduced cross-section method or the reduced property method can be used.

The situation outside Europe is different. The standard ASTM D 7247 (2007) was developed by a task committee, chaired by Borjen Yeh, which was formed by the North American engineered products industry in 2004. By this standard, adhesives that lose significant bond strength near the ignition temperature of wood should be screened out (Yeh et al., 2006). For acceptance, the adhesives must equal, or outperform, the adherend at a temperature of 230°C.

In the past, highly cross-linked phenolic and amino plastic adhesives dominated the bonding of load-bearing timber components. These adhesives are characterized by high strength and durability even at elevated temperatures. This performance information is based on decades of experience with these adhesives. A multitude of fire tests on glued laminated timber beams (Dorn and Enger, 1967; Dorn and Enger, 1967; Kolb, 1968; Dreyer, 1969; Kemmsies, 1999; Källander and Lind, 2006; Yang et al.,

2009; Frangi et al., 2011) led to the general belief that structural elements bonded with RF and PRF ensure thermal resistant bonding.

However, changes in customers' needs and innovations in adhesive technology brought in the last decade a number of new types of adhesives on the market that have faster curing at ambient temperatures, are free of water soluble or volatile organic compounds (VOC) and are applicable without mixing. Based on a lack of experience with these new types of adhesives, concerns have been raised regarding their performance at elevated temperatures.

According to the Adhesive Awareness Guide published by American Forest and Paper Association (2007), fire resistance tests by Forintek Canada and American Forest & Paper Association (AF&PA) have shown that the type of adhesive used can affect the fire resistance rating of end-jointed lumber used in stud wall assemblies. In these full-scale fire resistance tests, according to (ASTM E 119, 2011), PUR and PVAc adhesives reached a fire resistance rating of 51 and 49 minutes, respectively, compared to 1 hour in the case of PRF. The Glued Lumber Policy (American Lumber Standard Committee, 2009) generally differentiates between the types "Heat Resistant Adhesive (HRA)" and "Non-Heat Resistant Adhesive (Non-HRA)" on the grade stamp for the manufactured lumber. For acceptance as HRA, the adhesive is required to pass a one-hour fire resistance according to the standards ASTM D 7374 (2008) and ASTM D 7470 (2008), which require a wall assembly made with end-jointed lumber to be subjected to the ASTM E 119 (2011) fire test. The HRA mark permits the lumber to be used interchangeably with solid-sawn members of the same species and grade in fire-rated applications. Glued laminated timber beams with finger joints bonded with various structural PRF, MUF and PUR adhesives were also tested by (König, 2005). All tested adhesives complied with the current approval criteria for use in load-bearing timber components. In contrast to tests at ambient temperature, the beams for which PUR and MUF were used to bond the finger joints, exhibited bending resistances of only 70 to 80 % of the bending resistance of the beams with PRF-bonded finger joints.

Several investigations with small specimens free from defects at laboratory scale indicated that 1C PUR adhesives exhibit deficits regarding thermal stability. Shear tests of different adhesives at elevated temperatures by Frangi, Fontana and Mischler (2004) demonstrated that the behavior of PUR adhesives strongly depends on the formulation of the adhesive. Further publications of Richter and Steiger (2005); Richter, Pizzi and Despres (2006); Schrödter and Niemz (2006); Yeh et al. (2006); Niemz and Allenspach (2009) supported Frangi's assumption. The results showed that some of the 1C PUR adhesives were capable of attaining the thermal stability of PRF resins up to high temperatures, but others already failed below 100°C.

A statement by Dr. Jürgen König, specialist in fire safety for timber engineering about the topic is as follows:

"The use of some novel adhesives in bonded timber connections may imply reduced safety in fire compared to traditionally used PRF adhesives. The tests showed that beams bonded with some novel adhesives, both MUF and PUR, exhibit smaller fire resistance than those bonded with PRF, although

these adhesives are fulfilling current approval criteria in Europe. The North American approach – to exclude all adhesives not satisfying the performance requirements in small-scale tests with specimens pre-heated to a temperature of $225\,°C$ – is likely to exclude most novel structural adhesives that would perform well in a fire situation. This would counteract the wish of many countries to eliminate the use of formaldehyde. A change of the design codes, permitting the calculation of bonded connections in fire taking into account the thermo-mechanical material properties, would open the door for the use of novel, formaldehyde-free adhesives, would permit lower costs and reduce production time. Consequently, new classification for structural adhesives should be established with criteria with respect to their performance in fire." (König, Norén and Sterley, 2008)

In this research an approach different to König's suggestion is pursued. The objective of our research is, on the one hand, to investigate and, on the other, to improve the thermal stability of 1C PUR adhesives for structural applications to overcome prevailing deficits of the adhesives at elevated temperatures.

1.2 Motivation

Bonding of wood has been the focus of scientific research for many decades. Growing demands in architecture and civil engineering make improvements necessary for the future. The combination of ecologic and economic advantages of the material wood with contemporary design and innovative products and construction possibilities are the results of this trend.

1C PUR adhesives are widely used in the European wood engineering industry, since this type of adhesives has several advantages compared to conventionally used formaldehyde-based polycondensation adhesives. 1C PUR adhesives differ generally in their chemical structure as well as in their kinetics and therefore show a differing material behavior compared to competing systems.

Bonded wood joints significantly affect the material properties of structural elements, such as glulam beams or multilayer boards. In general adhesives, bonded joints and products for structural engineering have to meet high demands regarding strength and durability. These demands must also be guaranteed in the case of fire. Mechanical performance of wood and adhesives markedly decline with increasing temperature. Burn up tests on glulam beams showed that the load resistance of beams glued with 1C PUR or melamine was significantly lower compared to beams glued with PRF adhesive (König, Norén and Sterley, 2008).

The established formaldehyde-based condensation resins have been investigated in a multitude of scientific researches and have been further developed industrially. 1C PUR adhesives are at the beginning of their development and application and only limited scientific research has been done in this field so far. This is a main reason why adhesive producers and raw material suppliers support this project.

Research on polyurethane adhesives offers many starting points. Because of the great variety of raw materials (isocyanates and polyols) there are nearly endless possibilities in the chemical structure of

1 General Introduction

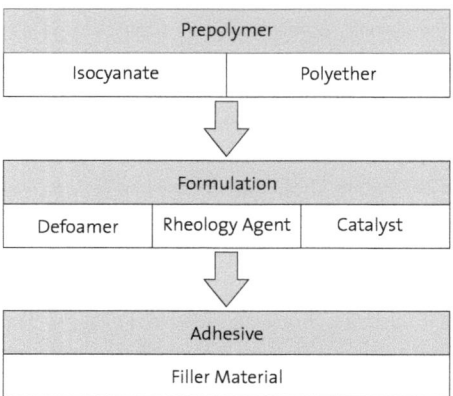

Fig. 1.3.1 Steps of adhesive development

polyurethanes. The chemical structure has great effects on the mechanical properties of the adhesive (Korley et al., 2006; Šebenik and Krajnc, 2007). Likewise the thermal stability of PUR depends largely on the chemical structure and is therefore different for each single configuration. This was shown in several investigations (Richter and Schirle, 2002; Richter and Steiger, 2005; Niemz, 2006; Richter, Pizzi and Despres, 2006; Schrödter and Niemz, 2006) by means of tests on bonded wood joints. Investigations on structural elements at high temperatures confirmed the huge differences in the results (Müller, 2000; Schmidt, 2001; Frangi, Fontana and Mischler, 2004; Frangi et al., 2011). However, the failure behavior of 1C PUR adhesives and their thermal stability have been insufficiently investigated and documented so far. Relationships between chemical structure and resulting material properties are not yet well established.

1.3 Main research objectives and arrangement of this thesis

With this thesis, PUR prepolymers and adhesive formulations were investigated to improve knowledge on the relationships between their chemical structure and mechanical, rheological and morphological properties. These structure-property relationships are analyzed relative to environmental, adherend and adhesive parameters. The thermal stability under mechanical load is of particular interest. The further development of the 1C PUR adhesives within this thesis is divided into three steps regarding the adhesive development, shown in Fig. 1.3.1.

In a first step, PUR prepolymers containing different types of isocyanates and polyether polyols were varied in regard to their chemical composition. In a second step, adhesives were formulated from the prepolymers. This meant adjustment of the kinetics (open time) and the rheology (viscosity) of the adhesive. Therefore different types of additives were added by the adhesive manufacturer Purbond in Sempach-Station, Switzerland. In the current work, an amine catalyst was used to accelerate the reaction of the prepolymers. The catalyst was added subsequent to the prepolymer production. Additionally, pyrogenic silica was used to adjust the rheology of the adhesive. Furthermore small amounts of defoamer and wetting agents were used to improve the processing of the adhesive. In a third step, different types of organic and inorganic filler materials were added to the formulations. Whereas some types of filler were added during the prepolymer synthesis, while others were added subsequently, during the formulation process.

1.4 Specific research objectives addressed in the papers

Paper I Several investigations were engaged with the thermal stability of adhesives at different temperature conditions. Transfer of results from single investigations may be difficult, since methods, adhesive formulations and environmental conditions vary. To gain an idea of the performance of commercially available adhesives this work focused on the comparison of the thermal-mechanical performance of different adhesive systems used in engineered wood products. The adhesives PRF, MUF, UF, EPI and PUR from different producers were included. Results obtained in this study are, in a manner of speaking, a benchmark for the adhesives that are developed in this project. Thereby the standard test method, DIN EN 302-1 (2004) for the determination of the bond strength under longitudinal tensile shear load, was used and applied at temperatures from 20 to 220°C.

Paper II The prepolymer, composed of isocyanate and polyol, is of central importance for the resulting product, since 1C PUR adhesives in general consist of more than 95 % of the prepolymer. Based on the great variety of raw materials it is possible to create products with a wide range of properties. Selective changes of important structural parameters makes it possible to study the effects they have on the adhesive properties. To minimize interactions between single parameters as much as possible, only one influencing factor should be varied at a time.

Paper III The remaining 5 % of the adhesive consist of additives that are admixed after the prepolymer production during the formulation process. These additives are first and foremost filler materials and in lower concentrations catalysts, defoamers, wetting agents, superplasticizers, dispersion mediums, etc. The effect of these additives in a standard formulation are the subject of the third paper, which investigates not only the macroscopic level, but also at the micro and nanoscopic level.

Paper IV Filler materials, which are the predominant additives, for adhesives make it possible to noticeably improve adhesive properties. However the storage stability and the rheological behavior of the adhesive are also affected by these substances. Formulations with inorganic filler materials were investigated to determine wether an improvement of the thermal stability of an adhesive is possible and how the filler material is distributed in the glue line area.

Paper V Based on their similar chemical constitution, organic materials are better suited as filler in combination with PUR. These materials are less likely to segregate and, ideally, are able to form covalent bonds with the prepolymer matrix. The aim of this last paper of the thesis was to investigate the potential of several organic filler materials to improve the thermal stability of 1C PUR adhesives. Therefore different prepolymer types and filler concentrations were tested at different temperatures up to 200°C.

2 Background to the adhesion of wood

2.1 Wood as an adherend

The material to be bonded – the adherend – has been pushed to the background within this thesis. This is because of the fact that it is hardly possible to adapt the naturally grown material wood to the adhesive. The only possibility would be selection. Properties such as density, elasticity, strength and wettability, which are all related to the anatomical structure and several wood species-specific features, are predicted by nature. A simple example of the different anatomical structure of hard and softwood is illustrated in Fig. 2.1.1.

There are certainly different possibilities to influence the glueability of wood by different mechanical surface treatments (Stehr, Gardner and Wålinder, 2001), plasma activating (Avramidis et al., 2010) or enzymatic treatments without (permanently) inserting additional substances. Using primers or coupling agents such as HMR (Vick and Okkonen, 1998; Vick and Okkonen, 2000; Christiansen,

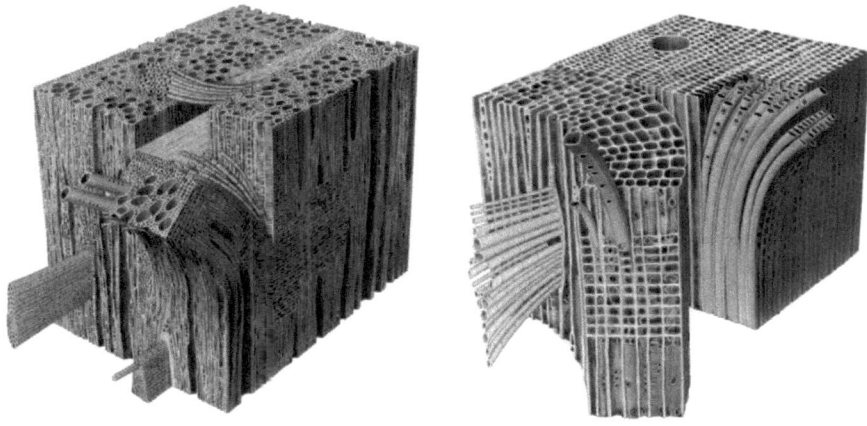

(a) Diffuse porous hardwood, beech (*Fagus sylvatica* L.)   (b) Softwood, Scots pine (*Pinus sylvestris* L.)

Fig. 2.1.1 Micro structure of wood (Bramwell, 1975)

2005; Follrich et al., 2007a) is another potential avenue of investigation that is worth enhancing to increase bonding effectiveness.

Wood is very sensitive to moisture and its mechanical properties therefore change remarkably under varying environmental conditions (Gerhards, 1982; Kudela, 1996; Szalai et al., 2004; Horvath, Molnar and Niemz, 2007). The influence of temperature on the mechanical properties is, within natural borders, limited, however at temperatures higher than 100°C the chemical bonds of wood start breaking. A significant pyrolysis of wood components starts from about 200°C (White and Dietenberger, 2001). Nevertheless, wood properties are also affected by low temperatures (Fengel, 1966; Schneider, 1971; Östman, 1985). Glos and Henrici (1991) determined, for engineered timber of spruce, a decrease of about 28 % in bending strength and of about 12 % in Young's modulus at 100°C compared to values obtained at room temperature. Kudela (1996) determined for beech wood a decrease of about 60 % in compression strength, however, it must be considered that the effect strongly depends on the moisture content.

2.2 Adhesion of wood

The adhesion of solid materials can be divided into three model assumptions; (1) mechanical adhesion (2) specific adhesion and (3) autohesion. The theory of autohesion of rubber elastic polymers is caused by diffusion of segments of chain molecules from one layer into another, consisting of the same material, and is of little relevance in the case of wood.

In the adhesion of wood, two models come into consideration: mechanical and specific adhesion. Mechanical adhesion describes a form-locking anchoring of the fluid phase of the adhesive within the surface of the substrate (Fig. 2.2.1). Since the anatomical structure of wood is based on a connected vascular system, it provides optimal conditions for penetration and anchoring of the adhesive. Independent of the load direction (perpendicular or parallel to the adherend surface), the mechanical adhesion can only contribute to a minor proportion of the adhesion but, in general, mechanical adhesion provides higher resistance to shear forces than to normal forces.

The bigger part of bonding energy has to be explained by specific adhesion theories. Under this term, several phenomena of chemical, physical and thermodynamical interactions are combined. Whereas chemical bonds play a subordinate role in wood adhesion, intermolecular forces, such as dipole forces, induction forces, dispersion forces or electrostatic forces, dominate. Several adhesion theories exist, which rely on these forces to explain the phenomenon of adhesion. These theories are dependant on the materials that are bonded, complementary to each other and to the mechanical adhesion.

Chemical bonding theory Covalent bonds between wood and adhesives have been controversially discussed by several authors, but firm verifications are pending or restricted to special environmental conditions or reactions of model resins that are not used in practice. Covalent bonds between the

2.2 Adhesion of wood

(a) Beech (*Fagus sylvatica* L.)   (b) Spruce (*Picea abies* Karst.)   (c) Douglas fir (*Pseudotsuga menziesii* Franco)

Fig. 2.2.1 Mechanical interlocking of the adhesion in different wood substrates

carboxylic acids of wood and PMDI are assumed by Wittmann (1976). Pizzi and Owens (1995) stated that, in the temperature range between 128 and 180°C, covalent bonding to wood components (lignin, cellulose and hemicellulose) is the predominant reaction in absence the of water. Pizzi, Mtsweni and Parsons (1994) also stated that covalent bonds between PF resin and lignin also appear to form but in a negligible amount regarding the overall adhesion. Very few publications about this issue suggest the existence of covalent bonds between PUR adhesives and wood (Schirle et al., 2002). Finally covalent bonding is of minor importance for wood adhesion.

Acid-base theory Originally developed by Brønstedt and Lowry in 1923, this theory was adopted to adhesion by Fowkes (1987). In this theory hydrogen bonds are considered as acid-base bonds whereas their bond energy depends on the acidity and basicity of the proton donator and acceptor, respectively.

Weak boundary layer (WBL) theory This theory is based on the assumption that the bond cannot fail exactly at the interface between adhesive and adherend (Robertson, 1975). The concept was developed by Bikermann (1967) and later adopted to wood adhesion by Stehr and Johansson (2000), who distinguished between a chemical (CWBL) and a mechanical weak boundary layer (MWBL). According to their publication, the main reasons for CWBL are extractives that migrate to the wood surface after machining. The MWBL is caused mainly by the machining operation itself or by surface degradation via natural ultraviolet radiation. On the one hand, the WBL forms a barrier for adhesives to penetrate into the adherend and restrains the mechanical adhesion. On the other, the damaged surface forms a zone of decreased strength properties.

Polarization theory According to De Bruyne and Houwink (1957), the theory is based on the dipole forces of molecules independent of their chemical compatibility. The forces are also described as physical bonds and can be divided into permanent dipoles (Keesom force), as occurs, for instance, in the water molecule due to the high proton number of the oxide atom compared to hydrogen atom.

The outcome of this is a higher electronegativity and the development of a dipole. These permanent dipoles have the ability to initiate forces to other dipoles or even induce a dipole (Debye force). Furthermore, there exist weak dispersion forces between non permanent dipolar molecules (London forces). Hydrogen bonds are a special type of dipole bond, developed by a hydrogen atom between two strong electronegative atoms, such as nitrogen, oxygen or fluorine. The interaction energy can amount to about 50 kJ mol^{-1}.

These types of bonds provide the most significant portion in the bonding of wood. In the case of polyurethanes, hydrogen bonding to the hydroxyl groups of wood can easily be developed by the carbonyl groups of urea or urethane linkages and hydroxy groups of wood, but also between NH groups of the polyurethane with oxygen atoms of the wood or adsorbed water molecules on the wood surface. Almost all wood components provide hydroxyl groups and lignin additionally provides carboxylic acid and ester groups for hydrogen bonding. A strong disadvantage of hydrogen bonds is their ability to be disrupted in the presence of water, however, this process is reversible. The cohesive strength of wood itself depends on hydrogen bonds and reveals, with increasing moisture content, decreasing values in stiffness and strength as a result of disrupted hydrogen bonds. After redrying, however, the original strength however is recovered. The same effect is observed for several adhesives including polyurethanes.

Electrostatic theory The theory developed by B. V. Derjagin in 1950 describes an electrical loaded zone, generated by charge transfers, with a thickness of several atoms at the border of two phases. This electrical double layer depends on charge carriers as electrons or ions, which are not present in wood.

Diffusion theory Voyutskii (1963) describes the mutual diffusion of polymer chains of adhesive and adherend and adhesive due to Brownian molecular movements of the macromolecules. For the adhesion of wood it would be more conceivable that the adhesive diffuses into the substrate rather than the opposite way. The diffusion theory also fits for the adhesion of wood, especially in areas that are swollen by the adhesive's solvent, since this condition facilitates the interaction between adherend and adhesive (Zeppenfeld and Grunwald, 2005). The mutual entanglement in the interphase region changes into a solid jointless bond, whereas a higher consistency improves the possible adhesion. The penetration of adhesives into the wood cell wall has to be classified on a higher structural level (Rapp et al., 1999; Gindl et al., 2004; Gindl et al., 2005; Konnerth and Gindl, 2006; Konnerth, Valla and Gindl, 2007; Konnerth et al., 2008).

Adsorption theory This thermodynamic theory of adhesion was developed by Sharpe and Schonhorn (1964) and is based on the wetting process of the adhesive to the substrate. The surface energy is associated with the contact angle of the wetting fluid. Due to the high surface roughness and occurring orienting and possibly chemical interactions between adhesive and wood, this theory is only

2.3 Wood properties affecting adhesion

2.3.1 Density of wood

Depending on the species, habitat, position in the stem, etc. the wood density varies significantly. Variations mostly result from different amounts of pore space and only marginally from differences in the amounts of the single wood components. The cell wall density of wood amounts to about $1500\,\mathrm{kg\,m^{-3}}$ as an overall mean for all species. For the assessment of the density influence on the adhesion, one has to differentiate between; (1) cohesive failure of the wood substrate, (2) cohesive failure of the interphase, (3) cohesive failure of the bond line or (4) adhesive failure. In the case of pure wood failure, there is a strong relationship between density of wood and its strength (Kollmann, 1951) since the adhered alone determines the strength of bonding. An increase in wood density hinders the adhesive in its penetration into the substrate, therefore reducing the mechanical interlocking of adhesive, yet protecting the bond line from starving. It is obvious that the latter is more important and therefore it is not amazing that the bond strength increases with increasing density (Widsten et al., 2006; Follrich et al., 2008). In the case of adhesion failure or cohesive failure in the bond line, the wood density has almost none to no influence on the adhesion (Haß et al., 2009).

2.3.2 Wood moisture and swelling and shrinking

One major function of the anatomical structure of wood is the transport of water. Soft and hardwoods differ in this regard, since softwoods realize water transportation and mechanical stabilization by the same cell type (tracheids). Hardwood, however, have a further cell type (vessels) exclusively for water transportation. This cell type forms greater diameters compared to tracheids. In green wood the water may be stored in the cell lumina. Below fiber saturation the water is only bound intracellularly by physical and chemical interaction with the cellulose and hemicellulose components and to some extent also with the lignin. The intracellular storage of water accounts for the swelling of the wood substrate. This process is caused by the hygroscopic behavior of wood depending on the relative humidity. Due to the fibrous structure of wood, a strong anisotropy of almost all properties is present. So even the swelling and shrinking is different in the anatomical directions whereas in the radial and tangential direction, 10 to 20 times greater values occur compared to the longitudinal direction. As an example, the differential swelling or shrinkage values for domestic species given in Niemz (1993) in the longitudinal direction are 0.01, in the radial direction 0.16 and in the tangential direction 0.32 % per % wood moisture change. Here the difference between soft and hardwood is rather small, whereas the maximum dimensional changes differ significantly.

2 Background to the adhesion of wood

Adhesives that are used for application under changing air humidity must be able to resist the internal stresses that occur with swelling and shrinking. The extent of these stresses is described, for example, by Gereke et al. (2009). This problem can be compensated by the use of ductile systems, such as PUR, that reduce occurring stresses, and also by brittle polycondensation resins, such as PRF, that resist high internal stresses. These water-based systems introduce additional stresses into the bond line due to surface swelling and subsequent shrinkage during reaction (Haß et al., 2012). The occurring stresses significantly increase under long time changing climatic conditions (Blanchet et al., 2003; Blanchet, 2008).

In the case of polyurethanes, the moisture content of wood is of special importance due to the chemical constitution of this type of adhesive. Since these adhesives need moisture for the cross-linking reaction, the wood moisture content must not drop below a certain value (Kägi, Niemz and Mandallaz, 2006). As green gluing shows, bonding is possible up to and beyond fiber saturation. In this process, the wood is glued directly after harvest (Na et al., 2005; St-Pierre et al., 2005; Sterley, Bümer and Wålinder, 2004; Pommier and Elbez, 2006; Sterley and Gustafson, 2006).

The moisture content is also important in regard to the chemical resistance of the adhesive against moisture. In the case of UF resin, contact with moisture, especially in combination with high temperatures, results in hydrolysis by generating free formaldehyde leading to the destruction of the bond line. MF, PRF, EPI or PUR adhesives however are considered as resistant against hydrolysis. A further problem of moisture is the disruption of hydrogen bonds, where the strongly polar water molecules substitute the bonding partners of the functional groups that participate in the hydrogen bond. Due to the disruption of hydrogen bonds, the adhesion and therefore the resistance against applied loads is decreased. The problem is also valid for intermolecular chain bonds whose disruption leads to a decrease of cohesion and a softening of the adhesive. This problem seems to be more pronounced for adhesives with lower chemical cross-linking such as PUR.

However, nanoindentation experiments by Konnerth et al. (2010) revealed a significant decrease in hardness and Young's modulus for polycondensation and reaction adhesives in wet state. It is remarkable that PUR and MUF reveal significantly lower decline compared to PRF and UF. In contrast to the polycondensation resins, PUR revealed nearly full recovery of stiffness after redrying. There is some indication that, in the case of polycondensation resins, bonds are irretrievably destroyed. The softening is also visible in the case of wood leading to a decrease of Young's modulus and strength with increasing wood moisture content. Therefore it is remarkable that PUR adhesives reveal significantly lower wood failure content if tested under wet conditions (Vick and Okkonen, 1998; Niemz and Allenspach, 2009).

The combination of high moisture exposure, internal stresses, mechanical loads and time often results in delamination of bonded wood joints. This effect is significantly increased for hardwood due to the higher maximum swelling values. Beech with red heartwood revealed even higher values in delamination tests (Aicher and Reinhardt, 2007). Both investigated systems, MUF and PUR, are able to resist delamination if the bonding is conducted under optimal conditions. By increasing the closed waiting time to at least 30 min, the delamination could be decreased to below 10 %, which is the criterium to pass the prevailing standard for adhesives used in structural applications DIN EN 301 (2006).

2.3.3 Orientation of the adherends

Based on the high anisotropy, the orientation of the adherends to each other is of special importance. Two characteristic angles are defined within the cartesian coordinate system, which does not fit to the anatomic structure of native wood, but instead to wood as a building material. The growth ring angle describes the orientation of the growth rings to the surface or in the case of bonding to the bond line plane. Zero degrees is the case if the radial orientation of wood is perpendicular to the bond line plane, 90 degrees if the tangential orientation is perpendicular to the bond line plane. In the case of beech used for bonding, investigations by Haß et al. (2009) showed that the relationship between the growth ring angle and the shear strength reveals the same tendencies as it does in plain wood as long as wood failure occurred; regarding the adhesion there is no significant effect. Results obtained by Furuno et al. (1983) for Hinoki (*Chamaecyparis obtusa* Endl.) led to the same conclusion.

The grain angle describes the angle between the longitudinal orientation of wood to the bond line. Zero degrees is the case if the bond line is parallel to the grain, and 90° in the special case of end grain joints. In practical applications the grain angle is below 5°, whereas the fibers are not aligned straight but rather in the shape of waves within the wood structure, partly branched or interrupted (Haß et al., 2010). Whereas Follrich et al. (2007b) revealed that the grain angle is of negligible importance to the shear strength, Świetliczny (1980) and Furuno et al. (1983) revealed significant differences regarding tensile shear strength for even small differences in the grain direction due to an anchoring effect of the penetrated adhesive into tracheid lumina, which also provides an increase of the surface area for bonding.

2.3.4 Chemical properties

The cell wall of wood contains several low-molecular organic (resins, fats, oils, terpens, tanins, sugars, acids, ketones, etc.) and inorganic (salts, sulfate, phosphate, etc.) compounds in the range of 5 to 30 % of the wood substance, depending on species, habitat, and time of harvest. These extractives on the one hand give wood its characteristic color and smell and on the other hand protect the living tree (and timber) against microorganisms and fungi. Beyond that exist metabolite intermediates of living cells, catalysts of herbal metabolism, energy suppliers and different forms of contaminations. Such wood extractives have a strong effect on the wettability and adhesion, which leads to a reduced bonding capacity (Wålinder and Johansson, 2001; Nussbaum and Sterley, 2002). Tests on several Australian wood species revealed the poorest gluability with PUR for species with relatively high bulk phenolic extractives and surface lipophilic extractives content (Widsten et al., 2006).

2.4 Penetration of adhesives – the interphase region

As a result of the porous structure, wood provides many pathways for adhesive penetration. The region of penetrated cells adjunct to the bond line is described as the interphase (Fig. 2.4.1(a)) and varies in size depending on anatomical, adhesive and processing parameters. Dominant cells for adhesive flow are vessels and longitudinal fibers (Fig. 2.4.1(b)) in hardwood and longitudinal tracheids (Fig. 2.4.1(c)) and rays in softwood species. The lumen diameters and the cumulation of these cells differ significantly between wood species and between the different cell types. Vessels of beech, for example, have a medium diameter of approximately 50 µm, whereas tracheids of spruce have a medium diameter of 10 µm. Since the penetration depends on the ratio between the average anatomical diameters of wood cells and the average molar mass distribution of the adhesive (Gavrilovic-Grmusa et al., 2010), it is not surprising that penetration into beech wood is much stronger compared to spruce. Depending on the molar mass of the adhesive system, penetration through interconnecting pits is also possible. Bordered pits hinder the adhesives between cells, whereas simple pits are only a minor obstacle (Gindl, 2001), however occlusions in the pits or lumens may inhibit flow.

The penetration of adhesives is mostly described by the penetration depth of the adhesive. In the case of urea-formaldehyde, the adhesive penetration into beech decreased with higher degrees of condensation of these resins. For this reason, (Schmidt, Glos and Wegener, 2010) increased the closed waiting time for bonding beech to prevent used MUF resin from penetrating too much. Haß et al. (2012) proposed a method to characterize the adhesive penetration as the saturation of the accessible pore space.

The penetration behavior of adhesives in the cellular structure has been investigated with several different methods described in a multitude of papers. On the majority, microscopic techniques (Suchsland, 1958; White, 1977; White, Ifju and Johnson, 1977; Johnson and Kamke, 1992; Johnson and Kamke, 1992; Rapp et al., 1999; Sernek, Resnik and Kamke, 1999; Wallström and Lindberg, 1999; Gindl, 2001; Rijckaert et al., 2001; Rijckaert, Stevens and van Acker, 2001; Gindl, Dessipri and Wimmer, 2002; Gindl et al., 2005; Kamke and Lee, 2007; Konnerth et al., 2008), mechanical techniques such as nanoindentation (Konnerth and Gindl, 2006; Konnerth, Valla and Gindl, 2007), porosimetry (Wang and Yan, 2005) and non-destructive, tomographic techniques where used (Niemz et al., 2004; Modzel, Kamke and De Carlo, 2011; Haß et al., 2012).

In order to investigate the causes of adhesive failure, a lot of research was done to determine the mechanical properties of the interphase region of the bond line. Appropriate methods like nanoindentation (Jakes et al., 2008) make it possible to determine the stiffness and hardness of the reacted adhesive systems directly in the bond line. Gindl, Schöberl and Jeronimidis (2004) proved the penetration of PRF resin into the cell wall of Norway spruce by comparing the mechanical properties of wood in relation to the distance from the bond line. Wood cells closer to the bond line revealed significantly higher hardness and stiffness. UV absorbance spectra of cell walls close to the bond line

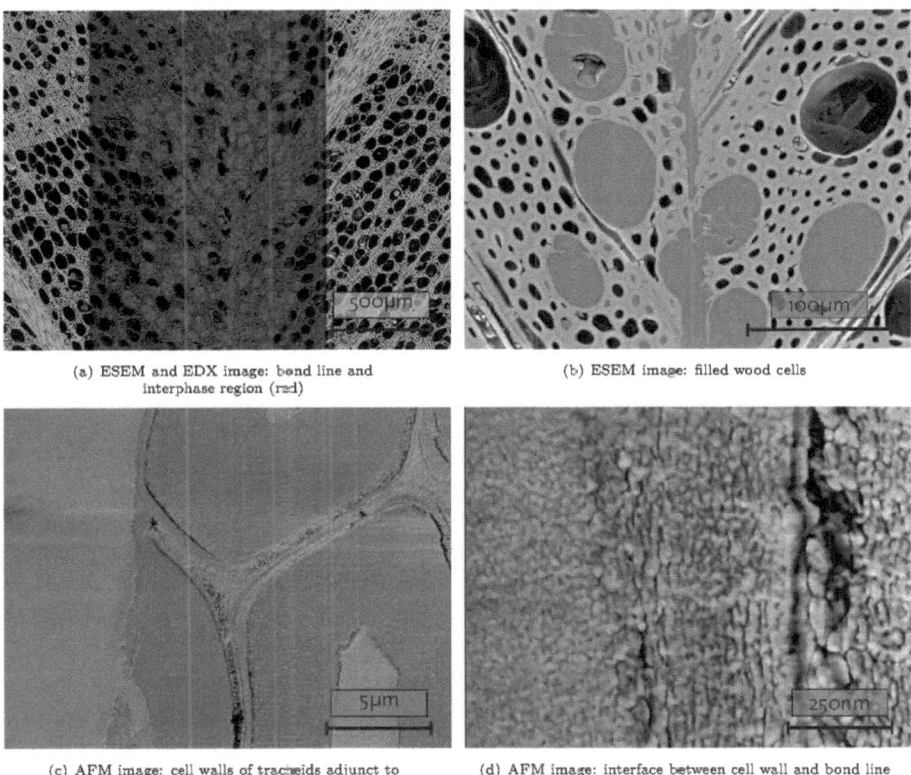

(a) ESEM and EDX image: bond line and interphase region (red)

(b) ESEM image: filled wood cells

(c) AFM image: cell walls of tracheids adjunct to the bond line

(d) AFM image: interface between cell wall and bond line

Fig. 2.4.1 Interphase region of the adhesive bond line in beech displayed on different size scales

also revealed different results compared to reference cell walls. PMDI in comparison did not show cell wall penetration. 1C PUR and PRF adhesives increased the stiffness in the interphase region due to penetration into cell cavities, whereas PRF revealed a greater impact compared to 1C PUR. (Gindl et al., 2005).

2.5 Chemistry of 1C PUR adhesives and specific properties affecting adhesion

Polyurethane chemistry goes back to the French chemist Charles-Adolphe Wurtz (1817-1884), who, in 1848, first described the reaction of monofunctional isocyanate with a monofunctional alcohol (Wurtz,

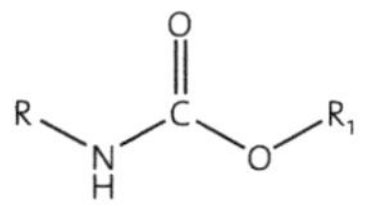

Fig. 2.5.1 Urethane group

1848) by forming the characteristic urethane group (Fig. 2.5.1). In this reaction the double bond between nitrogen and carbon is broken whereas the active hydrogen is bonded to the nitrogen and the oxygen to the carbon. A technical usage of (poly)urethanes became possible after an invention (DRP 728981) by Otto Bayer (1902-1989) in 1937 who described a process for the production of polyurethanes and polyureas (Bayer, 1947).

Polyurethanes can be found in a multitude of products and it is mostly used as foams, which are subdivided into hard, soft or integral-foams. The remaining part of cell-free products is used as solid materials, coatings, adhesives, fibers, microcapsules, gels and a lot more. For the production of polyurethanes aromatic isocyanates, such as isomeric mixtures of toluene diisocyanate (TDI), methylene diphenyl diisocyanate (MDI), as well as aliphatic isocyanates like hexamethylene diisocyanate (HDI), or isophorone diisocyanate (IPDI) (Fig. 2.5.2(a)-(d)), are used, whereas the aromatic compounds are preferred because of their much higher reactivity. Isocyanates are commonly produced by the reaction of the phosgene, formed by the reaction of chlorine with carbon monoxide, with at least one amine to form at least one isocyanate and hydrogen chloride. The polyol component mostly consists of polyether or polyester polyols produced from petrochemical raw materials, however, biogenous polyols based on vegetable oil are also possible (Somani et al., 2003).

The polyether polyols, which are the most commonly used products, build the backbone of the polyurethane. They result from a base-catalyzed reaction of bivalent or polyvalent alcohols with epoxides (ethylene oxide (EO) (Fig. 3.2.1(a)) and/or propylene oxide (PO) (Fig. 3.2.1(b)). By combinations of different epoxides, a large variety of long- and short-chained polyols with up to 8 OH groups per molecule and differing activity is available (Uhlig, 2006). The use of PO generates polyethers with secondary OH groups (Fig. 3.2.1(c)). The use of EO, on the other hand, causes more reactive primary OH groups when used at the end of the polyol synthesis. Due to a higher hydrophobicity of the resulting polyether, PO is preferentially used. Depending on the molar mass, the viscosity of the developed polyols exhibits a wide range. A further possibility is the reaction of diamines with epoxides to amino-polyether polyols. Polyester polyols are significantly more viscous and more expensive in their production compared to polyether polyols. Beyond that they are less stable against hydrolysis. Important advantages of polyurethanes made from polyester polyols is their photo-stability and a higher strength and stiffness caused by their greater potential for hydrogen bonding.

During the prepolymer production polyol mixtures with a surplus of polyisocyanate react to NCO-terminated polyether segments with an excess of free polyisocyanate. The cross-linking of the liquid polymer to a solid end product (Fig. 3.2.4) occurs in a moisture-curing reaction initiated by the

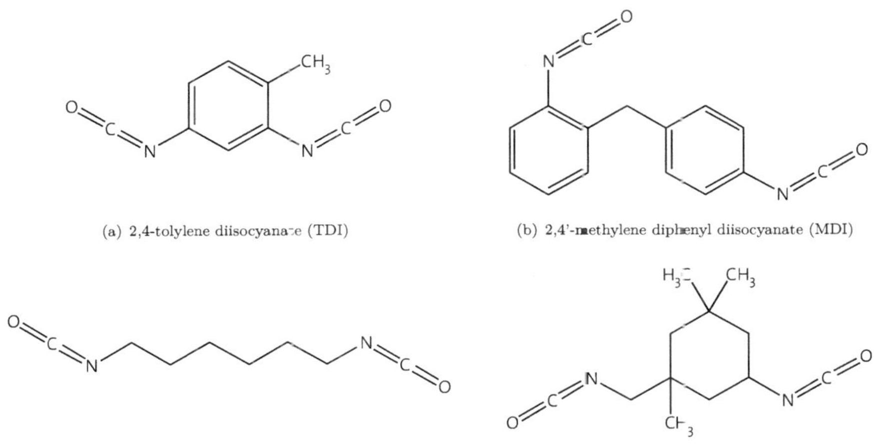

2.5 Chemistry of 1C PUR adhesives and specific properties affecting adhesion

(a) 2,4-tolylene diisocyanate (TDI)

(b) 2,4'-methylene diphenyl diisocyanate (MDI)

(c) 1,6-hexane diisocyanate (HDI)

(d) Isophorone diisocyanate (IPDI)

Fig. 2.5.2 Isocyanates for the production of PUR

ambient humidity and the water contained in the adherend. The reaction passes through an intermediate step in which carbamic acid is formed. The carbamic acid groups then dissociate to primary amines, which immediately react with further isocyanate groups to form polyurea groups. During the dissociation of carbamic acid, CO_2 is released, which results in foaming of the adhesive. A high NCO content and a fast reaction support this effect.

The cured polyurethane network is characterized by a segmented structure. The soft segments, consisting of flexible chains from the polyether oligomers, are joined by relatively rigid polyurethane-polyurea hard segments. The use of branched polyfunctional components ensures a high density of covalent cross-linking with higher thermal stability. The cross-linking of the hard segments is supported by hydrogen bonding of the NH groups and carbonyl groups of urea and urethane linkages. Due to their stronger hydrogen bonding, polyurea hard segments tend to agglomerate into hard segment domains so that the structure of the polyurethane partly becomes separated into hard and soft segment-rich phases. Hydrogen bonds are also formed between NH groups of urea and urethane linkages with ether oxygens from the polyether chains, however they are much weaker than those between hard segments. Secondary bonds exhibit a significantly lower bonding energy and thermal stability than primary bonds. In the polymer network, therefore, several structural bonding mechanisms interact with each other, thereby making it difficult to assign their individual contribution to a specific material behavior under temperature changes.

29

3 Main investigations

List of Papers

Paper I Thermal stability of glued wood joints measured by shear tests
Sebastian Clauß, Matúš Joščák, Peter Niemz
Eur. J. Wood Prod. (2011) 69:101–111

Paper II Influence of the chemical structure of PUR prepolymers on thermal stability
Sebastian Clauß, Dirk J. Dijkstra, Joseph Gabriel, Oliver Kläusler, Mathias Matner, Walter Meckel, Peter Niemz
Int. J. Adhes. Adhes. (2011) 31:513-523

Paper III Influence of the adhesive formulation on the mechanical properties and bonding performance of polyurethane prepolymers
Sebastian Clauß, Joseph Gabriel, Alexander Karbach, Mathias Matner, Peter Niemz
Holzforschung (2011) 65:835-844

Paper IV Improving the thermal stability of one-component polyurethane adhesives by adding filler material
Sebastian Clauß, Karin Allenspach, Joseph Gabriel, Peter Niemz
Wood Sci. Technol. (2011) 45:383-388

Paper V Influence of filler material on the thermal stability of 1C PUR adhesives
Sebastian Clauß, Dirk J. Dijkstra, Joseph Gabriel, Alexander Karbach, Mathias Matner, Walter Meckel, Peter Niemz
J. Appl. Polym. Sci. (2012) 124:3641-3649

3.1 Paper I

Eur. J. Wood Prod. (2011) 69:101–111

Thermal stability of glued wood joints measured by shear tests

Sebastian Clauß[1], Matúš Joščák[1] and Peter Niemz[1]

[1]Institute for Building Materials, ETH Zurich, Schafmattstrasse 6, 8093 Zurich, Switzerland

3.1.1 Abstract

The thermal stability of glued wood joints is an important criterion to determine the suitability of adhesives in the field of engineered wood. During their product life, glued wood joints can be exposed to high temperatures in various ways (direct exposure to the sun, fire, etc.). Thereby the cohesiveness of the adhesive must not degrade. This raises the question of how the strength of bonding changes under thermal load. The current investigation covers the influence of temperature (T=20 to 220°C) on the shear strength of glued wood joints. Different adhesive systems were investigated. With increasing temperature, the shear strength of solid wood and also of glued wood joints decreased. There were big differences in thermal stability and failure behavior between the adhesive systems as well as within the polyurethane group. The thermal stability of one-component polyurethane systems can be greatly varied by modifying their chemical structure. Well adapted one-component polyurethane adhesives reach a strength similar to that of phenol resorcinol resin.

3.1.2 Introduction

In practical use, glued wood components can be exposed to thermal loading in various ways. Behind glass facades of buildings, temperatures of about 60°C can be reached under direct exposure to the sun. These conditions can lead to failure of the supporting structure (Falkner and Teutsch, 2006). During fire, adhesives are exposed to even higher temperatures. In the outer regions of a timber beam, temperatures higher than 100°C can occur (Glos and Henrici, 1991). In contrast, much lower temperatures are measured inside the beams due to the poor thermal conduction of wood. Furthermore the water contained in the timber evaporates, so the introduced energy is partly converted into evaporation heat and delays an increase of temperature.

In the field of timber construction, the investigation of temperature influence on the adhesive performance has gained in importance over several years. Frangi, Fontana and Mischler (2004) investigated comparatively different types of adhesives. They found that a decrease in strength occurred over a

3 Main investigations

wide range of temperatures. Some of the tested one-component polyurethane (1C PUR) adhesives significantly lost strength from 70°C, others reached a good thermal stability up to high temperatures. Phenol-resorcinol-formaldehyde (PRF) resins showed an initial decrease of strength at around 180 to 190°C.

Investigations on the creep behavior of adhesive bonds were carried out by George et al. (2003) and Na et al. (2005), whereas a temperature dependant creep of PUR adhesives was found between 40 and 80°C. In the case of a higher initial strength (caused by a higher content of isocyanate), the creep in the low temperature range up to 50°C could be reduced. Richter, Pizzi and Despres (2006) investigated the relationship between the chemical structure and temperature-dependent creep properties of different commercial polyurethanes. They reasoned, by comparison of mechanical performance with ^{13}C-NMR spectroscopy, that the combination of a few chemical parameters had a big impact on the thermal stability of 1C PUR adhesives. These parameters were the relative proportion of remaining NCO groups in the polyurethane, the degree of polymerisation of the prepolymer and also the rate of reaction.

Within the current investigation, different commercially available adhesives have been investigated with respect to their thermal stability. The chosen adhesives are used in the wood industry and, with the exception of PVAc and UF, also in the field of engineered wood. A special focus was made on 1C PUR adhesives due to their controversially discussed behavior at high temperatures. A generalized conclusion for a type of adhesive by comparing different products without a variation of each type is not possible. The investigation rather gives an overview of currently used adhesives concerning their bonding strength in a wide range of temperatures.

3.1.3 Methods and materials

Wood

All bondings were carried out with beech wood (*Fagus sylvatica* L.). The raw density ρ at an EMC ω of $(13\pm1)\,\%$ amounted to $(756\pm54)\,\text{kg}\,\text{m}^{-3}$. The growth ring angle α (angle between growth rings and glued surface of the specimen) of the wood was between 30 and 90°.

Adhesives

For testing the thermal stability, commercially available adhesives (list below) from different producers were used. Table 3.1.1 lists the bonding parameters for each adhesive as they are recommended by the manufacturers and which were followed strictly. All used adhesives and hardeners were applied in liquid state.

- Urea-formaldehyde resin (UF)
- Melamine-formaldehyde resin (MF)

Table 3.1.1 Adhesives and their bonding properties

Adhesive	A/H[a]	DIN EN 301	EMC[b] [%]	p[b] [MPa]	t_p	T [°C]	Appl.	Spread [g m^{-2}]
PUR 1	-	✓	≥ 8	0.6-1.0	3 h	20	one side	200
PUR 2	-	✓	≥ 8	0.6-1.0	6.5 h	20	one side	180
PUR 3	-	✓	≥ 9	0.8-1.2	2.25 h	20	one side	250
UF	100/20	-	6-13	0.3-1.6	7 min	70	two sides	130
PVAc	-	-	n/a	0.8-1.2	15 min	20	two sides	150
MUF	100/35	✓	≈ 12	0.8-1.2	4 h	20	two sides	200
MF	100/10	✓	≈ 12	0.8-1.2	6 h	20	two sides	200
PRF	100/20	✓	≈ 12	0.8-1.2	5 h	20	two sides	180
EPI	100/15	✓	6-15	0.8-1.2	65 min	20	two sides	250

(A/H) adhesive/hardener ratio. (p) pressure, (t_p) pressing time. (T) temperature
[a] Applied in liquid state
[b] As recommended by adhesive producer

- Melamine-urea-formaldehyde resin (MUF)
- Phenol-resorcinol-formaldehyde resin (PRF)
- Polyvinyl acetate (PVAc)
- Emulsion-polymer-isocyanate (EPI)
- One-component polyurethane (1C PUR)

PVAc belongs to the group of thermoplastic polymers. These polymers are able to deform reversibly within a special temperature range. If this range is exceeded, a thermal degradation occurs. PVAc glues are, for example, used in the further processing of solid wood boards and veneers, and also in furniture manufacturing in general. The used adhesive was applied without hardener. For this reason, high thermal stability was not expected.

EPIs are reaction adhesives known for a relatively high thermal stability. By the cross-linking of polyvinyl alcohol with diphenylmethan-4,4'-diisocyanate (MDI) in combination with hydrophobic dispersions, high bonding strengths are achieved.

UF resins are the most common adhesives in the wood-based materials industry. Fields of application are particle boards, MDF boards, solid wood boards, plywood, engineered wood and furniture production. Due to their sensitivity to moisture (especially in combination with higher temperatures), UF adhesives often are fortified by cocondensation with melamine or phenol.

The tested MF adhesive is commonly used for bonded wood components (such as glulam) for supporting constructions. It is certified in combination with an appropriate hardener according to DIN EN 301 (2006).

MUF adhesives are produced either by separate production of UF and MF or by cocondensation of melamine, urea and formaldehyde in one and the same batch. The adhesives are characterized by a higher moisture stability compared to UF adhesives. They are used in the production of particle boards, MDF boards, solid wood boards, plywood and engineered wood.

PRF resins are generally used as cold curing adhesives for engineered wood products. Bondings with PRF are characterized by a good climatic stability and high bonding performance. From investigations on the temperature behavior of phenolic resins (PF), it is known that they tend to post-cure at high temperatures. Ohlmeyer (2003) showed that hot stacking of particle boards bonded with PF and MDI partly led to higher tensile strength perpendicular to the surface.

One-component polyurethane adhesives are classed as reactive adhesives. They are produced by a reaction of polyether polyols with a stoichiometric excess of isocyanate. Thereby long-chain polymers with isocyanate end and side groups are developed, which cure by a reaction with water contained in the wood. The isocyanate also reacts with the functional OH groups of the adherend. The used adhesives are fabricated by different producers and vary in their chemical composition. All of them are certified according to DIN EN 302-1 (2004) and DIN EN 301 (2006) standards and are primarily used in the field of engineered wood.

Production of the specimens

According to DIN EN 302-1 (2004), the prefabricated boards were stored under standard climatic conditions (20°C, 65 % RH) until the equilibrium wood moisture was reached. Directly before the bonding process, the boards were planed to the necessary thickness of (5±0.1) mm to exclude aging effects on the wood surface. The adherends were bonded with close contact bond lines ($\approx$0.1 mm) at room temperature and 50 % RH according to the producer's instructions in Table 3.1.1. The pressure for all adhesive bonds was 0.8 MPa. After one week storage under standard climatic conditions, the bonded adherends were cut into specimens according to DIN EN 302-1 (2004) (Fig. 3.1.1).

Testing procedure

The shear strength was determined according to DIN EN 302-1 (2004). To investigate the influence of the temperature on shear strength, 25 specimens of each group were tempered in a drying chamber for 1 h at 50, 70, 110, 150, 200 or 220°C, respectively. Subsequently, they were tested using an universal testing machine (Zwick Z100). The testing room of the machine was not tempered, thus the specimens' temperature could slightly decrease. Up to 70°C, the specimens were tempered in a plastic bag to ensure that the EMC remained constant. During the shear tests, the EMC corresponded to the wood moisture under standard climatic conditions. Above 110°C, the specimens reached oven-dry density. Reference specimens were stored under standard climatic conditions. The tests were performed position-controlled with a feed speed of 2 mm min^{-1}. The strain up to the maximal load was evaluated with a video-extensometer. After measuring the shear strength and strain, the wood failure percentage was estimated visually in 10 %-steps, as recommended in DIN EN 302-1 (2004).

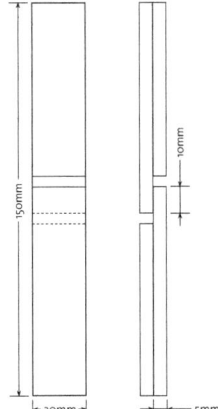

Fig. 3.1.1 Dimensions of test specimens according to DIN EN 302-1 (2004)

Furthermore, the specimens were weighed and the wood moisture content was determined according to ISO 3130 (1975).

3.1.4 Results and discussion

Solid wood specimens

To obtain a reference value for the strength of the glued specimens, non-glued solid wood specimens were tested in addition. In doing so, specimens were designed according to the geometry requirements of shear specimens in DIN EN 302-1 (2004). The geometry of the specimens and the direction of loading therefore differed from the requirements of the German standard DIN 52187 (1979) for wood specimens. Under standard climatic conditions, the shear strength $((15,0 \pm 3.5)$ MPa) corresponded to values determined by Horvath, Molnar and Niemz (2008). DIN EN 302-1 (2004) stipulates a growth ring angle of 30 to 90°; since this angle affects the shear strength of wood (Mikulja, Bogner and Zupcic, 2008), a higher variance occurred as a consequence of the orientation.

The shear strength of the tested solid wood specimens decreased with increasing temperature. At 110°C the shear strength dropped disproportionally and contradicts the linear trend which was shown by Frangi, Fontana and Mischler (2004) for the range between 20 and 170°C. For the bending strength Sonderegger and Niemz (2006) also showed a linear decrease with the temperature for different wooden materials in the range between -20 and 60°C. The same correlation is known from former investigations by Östman (1985) and Kudela (1996) for tensile and compressive strength, respectively.

The chemical components of wood undergo a thermal degradation which affects the strength properties if the material reaches elevated temperatures. From about 65°C, permanent reductions in strength are

3 Main investigations

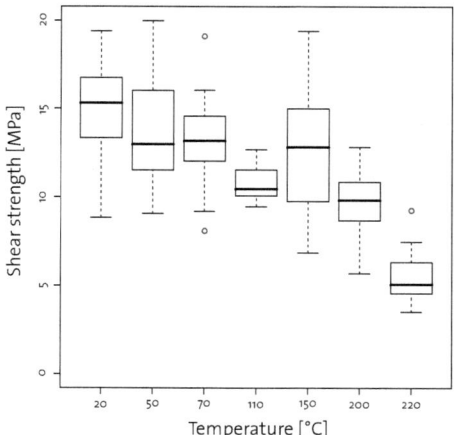

Fig. 3.1.2 Shear strength of beech against temperature (box: quartile of distribution, whisker: at most 1.5 × interquartile range, point: outliers)

possible. Thereby the amount of degradation depends on moisture content, temperature, exposure time, pH of wood and species. Reasons for strength degradation are depolymerisation reactions without significant weight loss. The chemical bonds of wood start breaking at temperatures higher than 100°C. Thereby lignin and carbohydrate weight losses occur that increase with the temperature. A significant pyrolysis of hemicellulose and lignin starts from about 200 and 225°C, respectively (White and Dietenberger, 2001). Probably the pyrolysis is the reason for the rapid reduction in strength between 200 and 220°C (Fig. 3.1.2). As a consequence tensile fracture was frequently found within this temperature range.

It should be pointed out that a temperature exposure in principle is accompanied by a change in wood moisture. Above 150°C, the average wood moisture of the specimens was 0 %. According to Horvath, Molnar and Niemz (2008), the reduction of wood moisture to less than 6 % leads likewise to a decrease in shear strength. The results therefore show an interaction of the effects.

Glued wood specimens

Table 3.1.2 shows the average shear strength of the tested adhesive bonds. Under standard climatic conditions, all adhesives passed the wood strength, which is shown by the high amount of wood failure. The polyurethanes PUR 1 and PUR 3 were also in the range of wood strength but the wood failure percentage was comparatively low (Table 3.1.3). With increasing temperature, PVAc proved itself the weakest adhesive. Around 50°C, the strength reached only 35 % of beech wood. However, PVAc must be considered as a special case as it was applied without hardener and is not certified for engineered

Table 3.1.2 Shear strength of adhesive bonds at variable temperatures

T [°C]						τ [MPa]					
		Beech	EPI	MF	MUF	PRF	PUR 1	PUR 2	PUR 3	PVAc	UF
20	$\bar{\tau}$	14.96	12.72	12.50	12.25	14.65	12.15	13.17	13.35	12.07	14.86
	s_τ	3.54	1.72	1.54	2.44	1.92	1.02	1.05	1.98	1.38	3.15
50	$\bar{\tau}$	13.90	11.70	11.81	12.54	14.95	8.94	11.76	11.30	4.90	13.39
	s_τ	3.36	1.12	1.17	2.47	2.08	1.31	1.05	0.98	1.31	1.55
70	$\bar{\tau}$	13.17	10.28	1.03	11.36	13.33	8.34	11.45	9.38	3.93	11.27
	s_τ	2.73	1.10	1.27	1.52	1.84	0.98	0.94	1.77	0.87	2.42
110	$\bar{\tau}$	10.89	10.82	0.53	8.75	10.86	9.56	11.86	10.87	3.90	11.48
	s_τ	1.03	1.72	1.84	2.46	1.51	1.55	2.00	1.22	1.43	2.45
150	$\bar{\tau}$	12.71	9.31	0.72	9.01	12.55	9.31	11.33	11.12	2.23	10.77
	s_τ	3.77	1.79	1.65	3.32	2.53	1.06	1.40	1.28	1.00	1.75
200	$\bar{\tau}$	9.57	7.46	8.80	9.11	10.1	7.94	9.57	8.56	0.96	6.59
	s_τ	1.96	0.86	1.54	2.76	1.88	1.79	1.38	1.69	0.51	2.41
220	$\bar{\tau}$	5.56	4.21	5.85	5.38	5.70	2.19	6.87	4.22	0.89	0.89
	s_τ	1.46	0.59	0.92	1.23	1.07	0.78	0.89	1.53	0.31	0.64

τ shear strength, $\bar{\tau}$ mean, s_τ standard deviation, T temperature

Table 3.1.3 Wood failure percentage of adhesive bonds at variable temperatures

T [°C]					Wood failure [%]				
	EPI	MF	MUF	PRF	PUR 1	PUR 2	PUR 3	PVAc	UF
20	90	100	100	90	40	90	20	80	100
50	100	100	100	90	10	90	0	0	80
70	70	100	100	90	10	90	0	0	70
110	90	100	100	100	20	90	70	0	100
150	90	100	100	100	20	100	40	0	60
200	40	100	100	100	40	100	40	0	70
220	40	100	70	100	0	90	20	0	0

wood construction. Up to 150°C, MF, PRF, PUR 2, PUR 3 and UF reached values beyond 80 % of the reference value. Around 220°C, only MF, PRF und PUR 2 exceeded the average of beech wood (with PUR 2 showing the highest shear strength). In this temperature range, EPI and UF also showed lower wood failure percentages.

Regarding the interpretation of these results, it should be considered that during the drying process a change in wood moisture content occurred. Up to 70°C, the wood moisture remained at about 12 % whereas it dropped to 0 % from temperatures of 150°C. Therefore, oven-dried specimens were tested. Fig. 3.1.3(a) shows the influence of temperature on the shear strength of all tested adhesives and beech wood. Between 70 and 150°C, the strength slightly increased, due to the reduction in wood moisture, which has a bigger effect on the strength than the increase of temperature (Kudela, 1996).

3 Main investigations

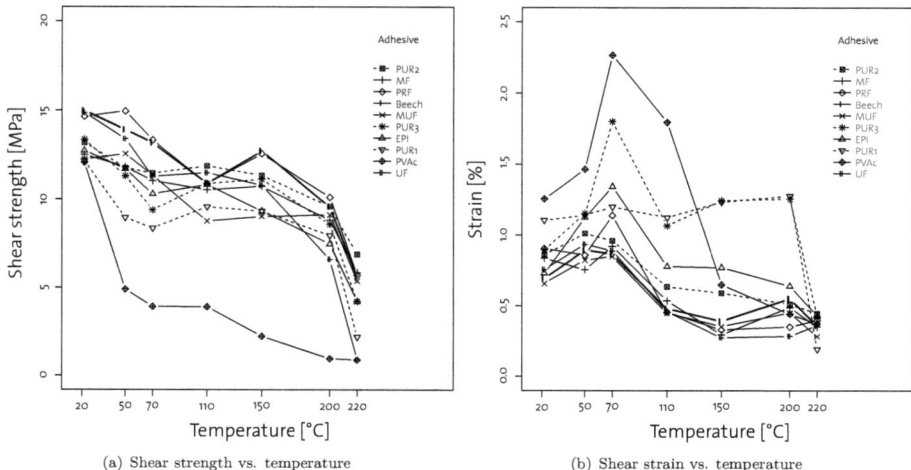

(a) Shear strength vs. temperature (b) Shear strain vs. temperature

Fig. 3.1.3 Results of tensile shear test of bonded wood joints vs. temperature

During the drying of wood and the corresponding shrinkage, residual stresses occur that interact with the external shear stress. These influences are hardly separable.

With regard to the strain of the bond lines, it is shown that the adhesives PVAc, PUR 1, PUR 3 and EPI caused higher strains in the whole temperature range, compared to the polycondensation adhesives (Fig. 3.1.3(b)). Maximum values for the shear strain were reached at a temperature of 70°C, which can be ascribed to the combination of increased temperature at a constant moisture content of about 12 %.

PVAc As shown in the box plot (Fig. 3.1.4(a)), the shear strength strongly decreased at 50°C. The wood failure percentage nearly dropped to 0 %. Due to the thermoplastic behavior of PVAc, higher strains occurred in the lower temperature range than for all other adhesives (Fig. 3.1.3(b)).

EPI Up to 150°C, EPI showed a good thermal stability (Fig. 3.1.4(b)). Compared to the reference specimens, EPI showed a 15 % lower strength throughout the whole temperature spectrum. Up to 150°C, the wood failure percentage was higher than 70 % and dropped below 50 % above this temperature. From about 200°C, it is assumed that the failure of the glued wood joints is caused by the adhesive. Due to the PVAc component, EPI shows a partly thermoplastic behavior, however it shows no obvious evidence for softening. From a chemical standpoint it is unclear if cross-links are destroyed or if the remaining thermoplastic behavior dominates.

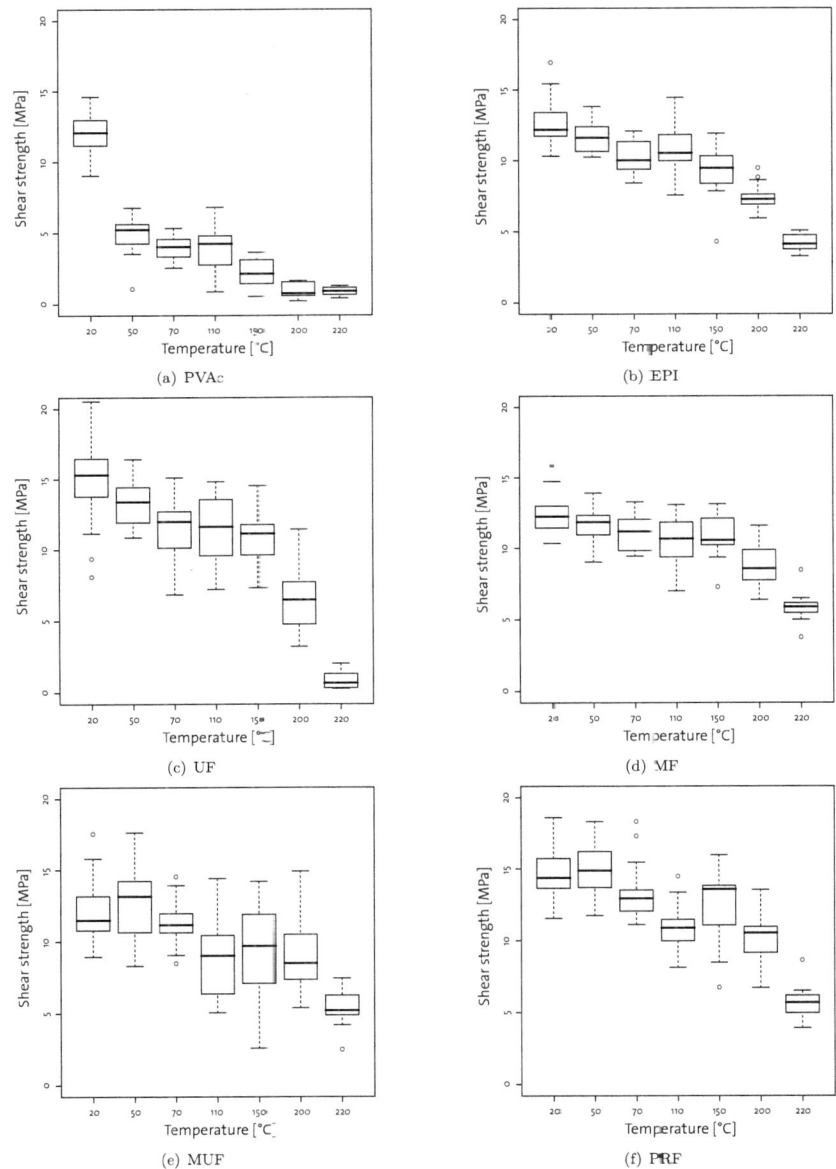

Fig. 3.1.4 Results of tensile shear test of bonded wood joints vs. temperature

3 Main investigations

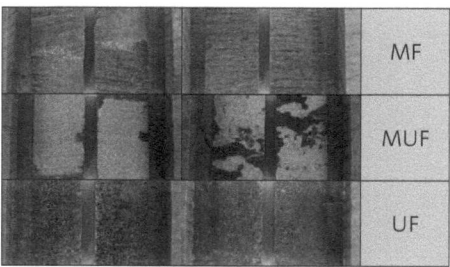

Fig. 3.1.5 Fracture surfaces of MUF, MF and UF at a temperature of 220°C

UF Up to 150°C, UF showed a high thermal stability (comparable with MF or MUF resins). In the case of lower temperatures, the adhesive reached just like PRF maximum values (Fig. 3.1.4(c)), which comply with the values of beech wood. At 200°C, the strength decreased more distinctly in comparison to the other adhesives. At 220°C the adhesive failed completely. Due to the short temperature exposure, the failure is probably not caused by hydrolysis effects. In contrast to MF, the fracture surface of UF (Fig. 3.1.5) shows a heavy discolouration of the adhesive caused by a thermal degradation.

MF The tested adhesive reached excellent shear strength at all temperatures (Fig. 3.1.4(d)). The wood failure percentage was nearly 100 % throughout the whole temperature range. It is clearly shown that the shear strength of the bond line was higher than the strength of beech wood. The images of the MF fracture surfaces (Fig. 3.1.5) show that even at 220°C no discolouration occurred and that wood fibres were pulled out all over the adherend.

MUF Up to 70°C, the bondings with MUF reached a shear strength above 10 MPa, which is the minimum shear strength according to DIN EN 301 (2006) under standard climatic conditions (classification type I). Fig. 3.1.4(e) shows a slight drop at 110°C to 9 MPa on average, which was below the values reached by MF und UF. In the temperature range above 150°C, the bonding strength was similar to the shear strength of beech wood. The wood failure percentage was up to 200°C 100 %. The lower shear strengths (compared to MF) at 110 and 150°C were caused by the failure of wood and not by the failure of the adhesive. The images of MUF fracture surfaces (Fig. 3.1.5) show like UF a heavy discolouration, but still high shear strength was observed. It can be argued that the degradation process (as indicated by the discolouration of the bond line) was delayed, due to the addition of melamin, whereas for UF this degradation process occurred at much higher rate.

PRF The bondings with PRF reached excellent shear strengths throughout the whole temperature range (Fig. 3.1.4(f)). Furthermore, nearly exclusively wood fracture occurred up to 220°C. The shear

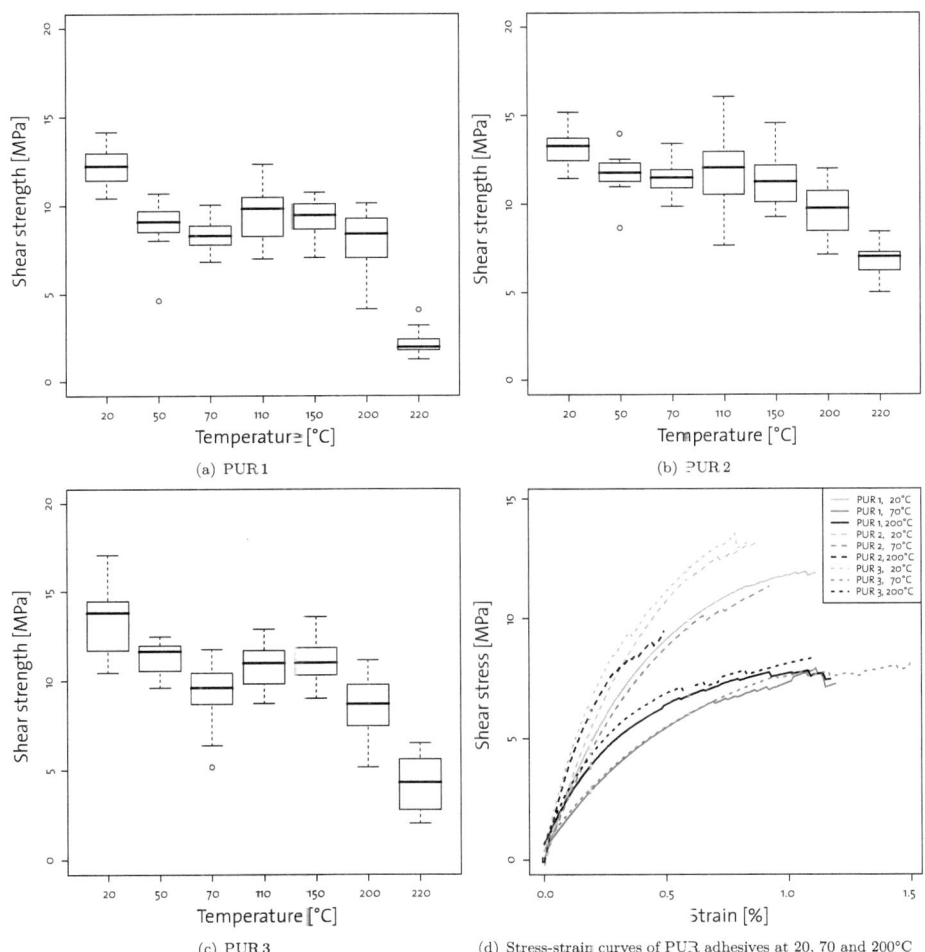

Fig. 3.1.6 Results of tensile shear test of bonded wood joints vs. temperature

strength exceeded the limit value of 10 MPa, according to DIN EN 301 (2006) under standard climatic conditions, up to 200°C.

1C PUR The 1C PUR adhesives from different producers showed diversity in their behavior. Under standard climatic conditions, all adhesives reached shear strengths between 12.0 and 13.5 MPa, which corresponds to 80 to 90 % of the shear strength of beech wood. Above 50°C, the strength of PUR 1 dropped significantly in comparison to the other PUR adhesives (Fig. 3.1.6(a)). The minimum shear strength, according to DIN EN 301 (2006) for classification type I, was reached only under standard climatic conditions. In the temperature range from 50 to 200°C, the adhesive withstood the temperature load at a shear strength above 7.5 MPa. At 220°C, the bonding suffered a clear decline compared to the other two 1C PUR adhesives. The wood failure percentage of PUR 1 was very low throughout the whole temperature range.

The PUR 2 reached, compared with the other PUR adhesives, the best results independent of temperature. Up to 150°C, the adhesive reached values significantly above 10 MPa (Fig. 3.1.6(b)) and thereby showed excellent thermal stability. Within the temperature range above 150°C, it also reached the highest overall shear strength. Above 200°C, the bonding reached a shear strength comparable to beech wood. The wood failure percentage was at least 90 % in the overall temperature range. This shows, in contrast to the other polyurethane adhesives, that primarily the wood material failed, even though at 20 and 50°C, nearly the same shear strength was reached compared to PUR 3.

The PUR 3 from a second producer reached high shear strengths as well, compared to the shear strength of beech wood (Fig. 3.1.6(c)). However, the adhesive showed significant disadvantages in the range between 70 and 110°C. The reason therefore might be found in the adhesive's chemical structure. At 220°C, the shear strength of the bonding dropped sharply. The wood failure percentage was considerably lower than that of the polycondensation adhesives. Furthermore, it turned out that the wood failure percentage increased at higher temperatures (Table 3.1.3).

The stress-stain-curves of the adhesives PUR 1 and PUR 3 (Fig. 3.1.6(d)) indicate a lower shear resistance at the temperatures 70 and 200°C, which is shown by the lower slope of this curves. The maximum strains of PUR 1 and PUR 3 at these temperatures (especially PUR 3 at 70°C) were also increased, compared to PUR 2. However, PUR 2 showed a higher maximum strain than beech wood as well (Fig. 3.1.3(b)). In the range of low deformation ($\varepsilon \leq 0.5\,\%$), the slope of PUR 2 remained nearly constant independent of temperature. That confirms the higher thermal stability of this adhesive. The reduced stiffness of two of the polyurethanes is possibly caused by a plasticizing effect due to the coaction of the remaining wood moisture content and the elevated temperature. To confirm this assumption, further investigation is needed.

3.1.5 Summary and conclusion

In the lower temperature range, all adhesives showed sufficient shear strengths above 10 MPa (Fig. 3.1.7(a)). The best results compared to the average wood strength were reached by PRF and UF adhesives. The wood failure percentage reached at least 80 %, with the exception of PUR 1 and PUR 3.

In the range between 50 and 150°C, the adhesives showed good thermal stability. Only PVAc failed at 50°C due to its thermoplastic behavior. Thereby it must be pointed out that temperatures up to

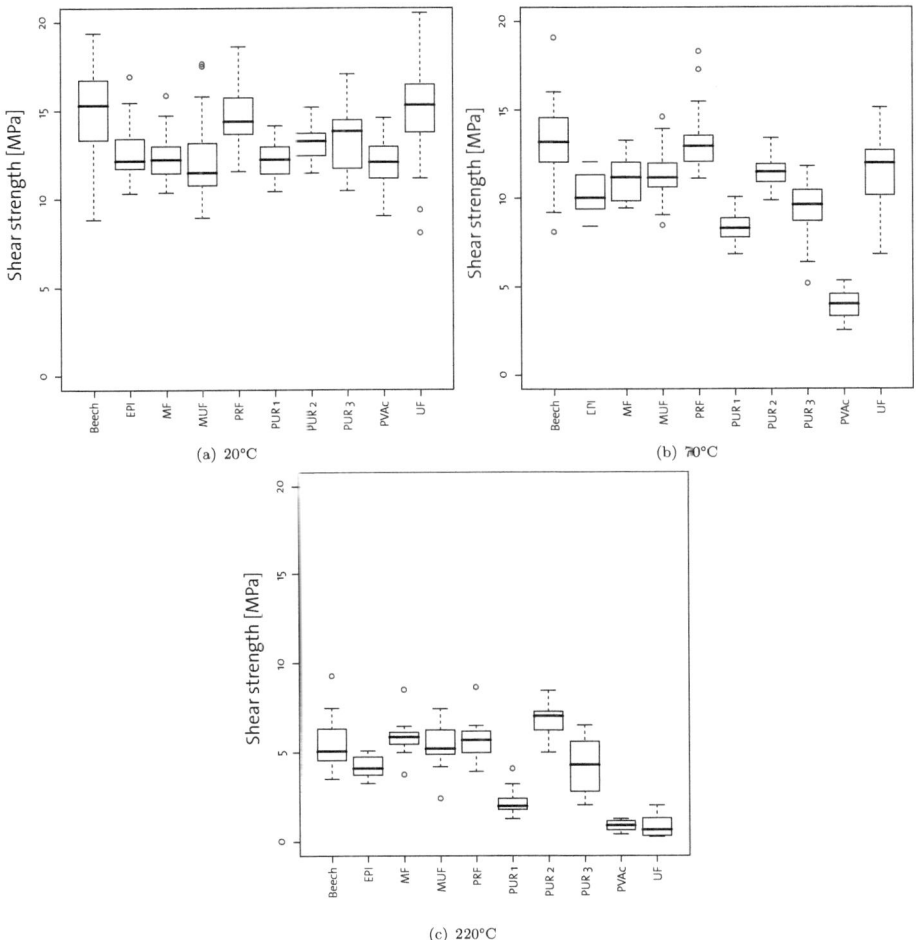

Fig. 3.1.7 Comparison of tensile shear strength of adhesive bonds at selected Temperatures

70°C are in a range of practical relevance. From about 70°C (Fig. 3.1.7(b)), two of the polyurethane adhesives showed a growing decrease in strength. The wood failure percentage of these adhesives (PUR 1, 20 %; PUR 3, 30 %) also differed significantly from that of PUR 2 and the polycondensation resins. A possible reason for this behavior is the different reactivity of the PUR adhesives. PUR 1 (60 min) and PUR 3 (45 min) have up to half the open time compared to PUR 2 (120 min). Furthermore, the viscosity of the adhesives increases from PUR 1 (3000 mPas) to PUR 3 (12000 mPas) to PUR 2 (17000 mPas). Since exact information about the chemical structure of the adhesives is not known to the authors, more detailed research is necessary to explain the observed effects.

At 200°C, PUR 1 and UF showed a decreased thermal stability. The shear strengths of these adhesives differed significantly from the rest. The decrease in strength was also reflected in the wood failure percentage, which descended to 0 %. Compared to the average wood strength of all tested adhesives, also EPI and PUR 3 slightly decreased in strength. Up to 220°C, MF, MUF, PUR 2 and PRF showed mostly wood failure and reached shear strengths similar to beech wood (Fig. 3.1.7(c)). Thereby it must be considered that in case of PRF, MUF, MF and PUR 2 not the adhesive but rather the wood properties were testet. Thus a comparison of the adhesives' properties is not feasible. To deal with this problem and to get more information about the adhesive properties Konnerth et al. (2006) used a different specimen geometry. Pizzo et al. (2003) on the other hand used a different testing set up to compare the glue line and the solid wood with the same specimen. This procedure offered the advantages that a direct comparison becomes possible and the influence of wood variability is minimized. For the practical relevance, it is only sufficient to exceed specific values which guarantee a suitable strength.

The 1C PUR adhesives showed a high variation within their adhesive group. This also applies for the shear strength and the wood failure percentage. While PUR 1 only reached a shear strength above 10 MPa at 20°C, PUR 2 reached values comparable to beech wood up to 220°C. In conclusion, 1C PUR adhesives exhibit a wide range of properties caused by their spectrum of assembly possibilities.

Apart from the temperature, some other influencing factors, such as wood moisture and raw density, were present under the applied experimental circumstances. Their impact on the results must be considered. Up to 70°C, the moisture content was approx. 12 %. By contrast, above 110°C the specimens were oven-dried. Due to the change in wood moisture and the corresponding shrinkage of wood, residual stresses occur and lead to an additional decrease of shear strength. Apart from the bonding, the wood moisture itself influences the shear strength (Schrödter and Niemz, 2006). Investigations on beech wood (Horvath, Molnar and Niemz, 2008) revealed an increasing compressive shear strength beginning from oven-dried specimens up to 6 % wood moisture. By contrast, with a further increase of the moisture content, the compressive shear strength decreased. This fact makes it difficult to derive a dependency of shear strength on the temperature. The influencing factors are difficult to separate over such a wide range of temperatures. A possible way could be to oven-dry the specimens and to test them subsequently. By changing this experimental procedure, the influence of wood moisture can be eliminated but the initial drying may cause damage to the specimens.

3.1.6 Acknowledgements

The authors would like to thank for many helpful suggestions given by Dr. J. Gabriel (Purbond, Sempach-Station, Switzerland).

References

DIN 52187 (1979) Testing of wood – determination of ultimate shearing stress parallel to grain. Berlin: Beuth Verlag

DIN EN 301 (2006) Adhesives, phenolic and aminoplastic, for load-bearing timber structures – Classification and performance requirements. Berlin: Beuth Verlag

DIN EN 302-1 (2004) Adhesives for load-bearing timber structures – Test methods – Part 1: Determination of bond strength in longitudinal tensile shear strength. Berlin: Beuth Verlag

Falkner, H. and **Teutsch, M.** (2006) Load-carrying capacity of glued laminated wood girders under temperature influence. Bautechnik, 83 No. 6, 391–393

Frangi, A., Fontana, A. and **Mischler, A.** (2004) Shear behaviour of bond lines in glued laminated timber beams at high temperatures. Wood Sci. Technol. 33 No. 2, 119–126

George, B. et al. (2003) Comparative creep characteristics of structural glulam wood adhesives. Holz Roh Werkst. 61 No. 1, 79–80

Glos, P. and **Henrici, D.** (1991) Bending strength and MOE of structural timber (*Picea abies*) at temperatures up to 150°C Holz Roh Werkst. 49 No. 11, 417–422

Horvath, N., Molnar, S. and **Niemz, P.** (2008) Examinations to the influence of wood moisture on chosen wood properties of spruce, oak and beech. Holztechnologie. 49 No. 1, 10–15

ISO 3130 (1975) Wood – Determination of moisture content for physical and mechanical tests. Berlin: Beuth Verlag

Konnerth, J. et al. (2006) Comparing dry bond strength of spruce and beech wood glued with different adhesives by means of scarf- and lap joint testing method. Holz Roh Werkst. 64 No. 4, 269–271

Kudela, J. (1996) Influence of moisture and temperature loading on strength of beech wood loaded in compression. Drev. Vysk. 41 No. 2, 3–17

Mihulja, G., Bogner, A. and **Zupcic, I.** (2008) Gluing strength of wood measured with nonstandard pressure-shear method. Wood Res. 53 No. 1, 91–104

Na, B. et al. (2005) One-component polyurethane adhesives for green wood gluing: Structure and temperature-dependent creep. J. Appl. Polym. Sci. 96 No. 4, 1231–1243

Ohlmeyer, M. (2003) Improvement of panel properties by hot stacking. In 37th international wood composite materials symposium proceedings. Washington State Univ Pullman WA, 109–118

Östman, B.A.-L. (1985) Wood tensile strength at temperatures and moisture contents simulating fire conditions. Wood Sci. Technol. 62 No. 6, 424–432

Pizzo, B. et al. (2003) Measuring the shear strength ratio of glued joints within the same specimen. Holz Roh Werkst. 61 No. 4, 273–280

Richter, K., Pizzi, A. and Despres, A. (2006) Thermal stability of structural one-component polyurethane adhesives for wood - Structure-property relationship. J. Appl. Polym. Sci. 102 No. 1, 24–32

Schrödter, A. and Niemz, P. (2006) Investigation on the failure behaviour of glue joints at high temperatures and relative humidity. Holztechnologie, 47 No. 1, 24–32

Sonderegger, W. and Niemz, P. (2006) The influence of the temperature on the bending strength and the modulus of elasticity of diverse wooden materials. Holz Roh Werkst. 64 No. 5, 385–391

White, R.H. and Dietenberger, M.A.; Buschow, K.H.J. et al., editors (2001) Chap. Wood Products: Thermal degradation and fire. In Encyclopedia of materials: Science and technology. 2nd edition. Oxford: Elsevier Science Ltd., 9712–9716

3.2 Paper II

In: J. Adhes. Adhes. (2011) 31:513-523

Influence of the chemical structure of PUR prepolymers on thermal stability

Sebastian Clauß[1], Dirk J. Dijkstra[2], Joseph Gabriel[3], Oliver Kläusler[1], Mathias Matner[2], Walter Meckel[3], Peter Niemz[1]

[1] Institute for Building Materials, ETH Zurich, Schafmattstrasse 6, 8093 Zurich, Switzerland
[2] Bayer MaterialScience AG, Coatings, Adhesives & Specialties, Kaiser-Wilhelm-Allee 1, 51373 Leverkusen, Germany
[3] Purbond AG, Industriestrasse 17a, 6203 Sempach-Station, Switzerland

3.2.1 Abstract

The thermal stability of adhesives for load-bearing construction has been one of their key parameters since engineered wood products were introduced in timber construction. In the case of one-component moisture-curing polyurethane (1C PUR) adhesives, knowledge about relationships between their chemical structure and the resulting bonding properties is limited, especially under high-temperature conditions. In this study the structure-property relationships of 1C PUR prepolymers were analyzed in the temperature range from 20 to 200°C by means of mechanical and rheological tests. NCO-terminated urethane prepolymers were prepared from systematically varied MDI and polyether mixtures. The structural parameters investigated were the urea and urethane group content, cross-link density, ethylene oxide content and the adjustment of functionality via NCO or polyether component. Bonded wood joints were tested for their tensile shear strength and polymer films were analyzed by means of DMA and DSC. The results revealed a significant influence of hard segment content and cross-link density on the thermal stability of the prepolymers. Not all parameters that affect the film properties significantly influence bonding.

3.2.2 Introduction

Fire resistance and thermal stability are still among the important criteria for engineered wood products applied in building construction. Since this commonly concerns glued wood products, adhesives also become a crucial link in the chain. In the past, highly cross-linked phenol-resorcinol-formaldehyde (PRF) adhesives dominated this market segment, but changes in customer needs and adhesive technology brought new types of adhesives; one example is one-component moisture-curing polyurethane (1C PUR).

Thereupon, concerns arose that adhesives with a reduced cross-link density perform worse at elevated temperatures. Several investigations indicated that 1C PUR adhesives, generally made from polyether polyols and an excess of polyisocyanates, exhibit deficits regarding thermal stability (Properzi et al., 2002; George et al., 2003; Frangi, Fontana and Mischler, 2004; Na et al., 2005; Richter and Steiger, 2005; Richter, Pizzi and Despres, 2006; Richter et al., 2006; Yeh et al., 2006). In particular, temperature-dependent creep is deemed to be the main disadvantage of this adhesive type (Properzi et al., 2002; George et al., 2003; Na et al., 2005; Richter and Steiger, 2005; Richter, Pizzi and Despres, 2006; Richter et al., 2006).

Studies showed a sharp decline in stiffness with increasing temperature from 50 to 100°C (Properzi et al., 2002; George et al., 2003; Na et al., 2005; Richter and Steiger, 2005; Richter et al., 2006), which is definitely characteristic for semi-crystalline thermoplastics with a glass transition temperature below room temperature such as polyurethane. However, they do not describe creep in a physical sense, implying a time- and temperature-dependent deformation of the material under constant load. The decrease in storage modulus with increasing temperature indicates an increase in polymer chain flexibility due to the disappearance of physical bonds above about 110°C (Woods, 1990). The temperature-dependent decrease in the adhesive bond strength, which is more relevant from the practical point of view, was studied at temperatures up to 230°C for commercial and non-commercial polyurethane adhesives and competitive polycondensation adhesives (Frangi, Fontana and Mischler, 2004; Schrödter and Niemz, 2006; Yeh et al., 2006; Niemz and Allenspach, 2009; Clauß, Allenspach and Niemz, 2011; Clauß, Joščák and Niemz, 2011). The results showed that some of the 1C PUR adhesives were capable of attaining the thermal stability of PRF resins up to high temperatures but others already failed below 100°C.

A drawback of studies on commercially available adhesives is their largely unknown formulation. To obtain information about the chemical structure, ^{13}C-NMR spectra were collected from the resins (Na et al., 2005; Richter et al., 2006) and correlated with their mechanical performance, and clearly a higher or lower isocyanate-to-polyol ratio did not affect the temperature-dependent storage modulus (Na et al., 2005). According to further publications, reported in Šebenik and Krajnc (2007) and Chattopadhyay and Webster (2009), the NCO/OH ratio and the molar mass are important factors determining the structure and thermal properties of PUR. The results in (Na et al., 2005) further revealed a higher stiffness at lower temperatures (up to 50°C) as a consequence of a higher isocyanate proportion. Additional investigations on the laboratory scale, using combinations of different isocyanates and amines, showed that higher hard segment proportions and high cross-linking yielded better adhesive performance. Furthermore it is reported that a higher proportion of still reactive, free NCO groups, a lower degree of resin polymerization and a slower reaction rate are the dominant factors for high thermal stability (Richter et al., 2006), whereas beyond a critical free NCO content, the high stiffness of the adhesive is responsible for a decrease in adhesion strength (Šebenik and Krajnc, 2007).

To ensure that modern adhesives fullfill the currently more demanding requirements regarding thermal stability, building code agencies, research institutes and the industry established new standards to test adhesives for engineered wood products at temperatures up to 230°C. The US standard ASTM D 7247 (2007), approved and published by the ASTM, is one example. According to this, the acceptance criterion for adhesives is to be equal to or even outperform the adherend. Another test (DIN EN 14292, 2005), developed by EMPA in Switzerland in 2005 and approved by the German Institute of Standardization, is aimed at estimating the time to rupture at a constant temperature rate of 50°C per hour. A third standard to evaluate the thermal stability of adhesives (DIN EN 14257, 2006) is geared to the standard test method to determine the bond strength of adhesives for load-bearing constructions (DIN EN 302-1, 2004) at an elevated temperature of about 80°C.

The situation during these tests, however, is not comparable to the situation in the bond line of laminated timber during burn-up. Wood itself is a very good insulator ($\lambda_{(Spruce, T=20°C\ \omega=12\%, radial)} = 0.086\,W(m\,K)^{-1}$ (Sonderegger, Hering and Niemz, 2011)) and exposed to fire it forms a surface char layer with very low temperature-dependent thermal conductivity. This layer protects the timber interior against heat. Due to this good insulating behavior, temperature profiles across wood members exposed to fire exhibit a steep gradient. Temperatures of about 230°C as used in ASTM D 7247 (2007) occur only very close to the char front. The thermal penetration depth (the distance between the char-line and the part of the wood at room temperature) is in the range of 25 to 50 mm (Frangi and Fontana, 2003). According to calculation models presented by Frangi and Fontana (2003), after 30 min ISO fire, at a distance of 35 mm from the original surface, the temperature is already below 100°C. Consequently, the whole range from room to ignition temperature is of particular interest if the behavior of a laminated timber construction is to be described in the case of fire.

The thermal stability of polyurethanes and the factors influencing their mechanical performance have been extensively studied in recent years. All these investigations revealed a big variation in the performance of 1C PUR adhesives, but firm statistical results, regarding structure-property relationships, have not been provided. In the present study, the chemical structure of 1C PUR prepolymers was systematically changed. In doing so, the isocyanate (NCO) content, the cross-link density, the urethane group content and the ethylene oxide/propylene oxide ratio were varied. Furthermore it was studied if there is a difference whether the functionality is given by the isocyanate or the polyether component of the polymer, respectively.

To investigate the temperature-dependent bonding performance of these laboratory prepolymers, different approaches were used. The temperature in the tests ranged up to 200°C. On the one hand, tensile shear tests were carried out to determine the bonding strength of wood joints. On the other hand, the viscoelastic nature of the polymers was tested by means of dynamic mechanical analysis of the polymeric films.

3 Main investigations

3.2.3 Material and methods

Prepolymers

The prepolymers analyzed in this study result from the reaction of a polyol component with a surplus of a polyisocyanate component. They represent the first step of cross-linking as a reactive intermediate between monomeric isocyanates and polyurethane polymers.

The polyol results from a base-catalyzed reaction of bivalent or polyvalent alcohols (1,2-propylenglycol, glycerol and trimethylolpropane (TMP)) with epoxides (ethylene (EO) (Fig. 3.2.1(a)) and/or propylene (PO) oxide (Fig. 3.2.1(b))). Due to the combination diversity, a large range of polyether polyols with varying functionality is available. Using PO generates polyethers with secondary OH groups (Fig. 3.2.1(c)); EO, on the other hand, causes more reactive primary OH groups when used at the end of the polyol synthesis. Due to a higher hydrophobicity of the resulting polyether, preferentially PO is used. The reactivity of the polyethers with NCO groups can be easily adjusted by varying the ratio of the epoxides. Depending on the molar mass, the viscosity of the developed polyols exhibits a wide range.

The isocyanate component consists of a mixture of methylene diphenyl diisocyanate (MDI) isomers (Fig. 3.2.2) where among the highly reactive 4-4'-MDI mainly polymer MDI with a functionality > 2 is contained.

While stirring constantly, the dry polyol component is mixed with the NCO component at a temperature of between 50 and 70°C until the constant NCO content targeted is reached. During the polyaddition reaction between the two components (Fig. 3.2.3), polyurethane oligomers with highly reactive NCO end groups are formed. Additionally, a surplus of unreacted monomeric isocyanate remains.

The cross-linking of the liquid polymer to a solid end product (Fig. 3.2.4) occurs in a moisture-curing reaction initiated by the ambient humidity and the water contained in the adherend. The reaction

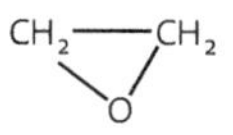

(a) Ethylene oxide

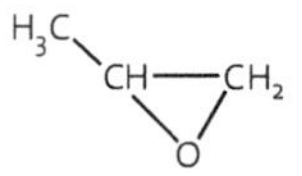

(b) Propylene oxide

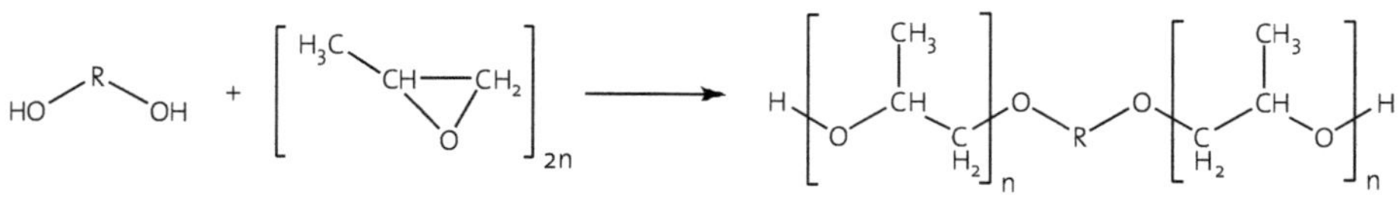

(c) Reaction of bivalent start alcohol and propylene oxide to polypropylene glycol

Fig. 3.2.1 Epoxides and their reaction with alcohol to polyether

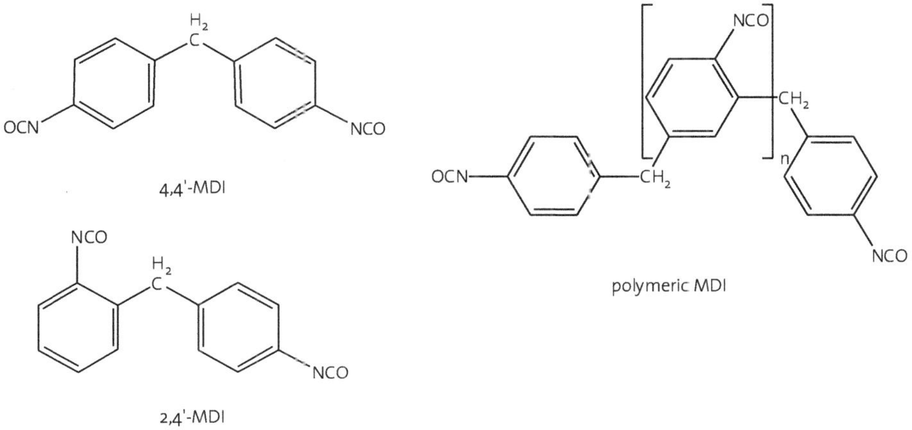

Fig. 3.2.2 Methylene diphenyl-4-4'-diisocyanate (4,4'-MDI), 2,4'-MDI and polymeric MDI

passes through an intermediate step in which carbamic acid is formed. The carbamic acid groups then dissociate to primary amines which immediately react with further isocyanate groups to form polyurea groups. During the dissociation of carbamic acid, CO_2 is released, which results in foaming of the adhesive. A high NCO content and a fast reaction support this effect.

The reaction of water with isocyanate groups is a very clean reaction that results in an almost 100 % formation of urea groups, representing one of the best hard segments in PUR chemistry giving strong physical cross-linking via hydrogen bonding. Only at the very end of the curing may formation of free amine groups occur due to a lack of mobility of the polymeric chains. Other possible side reactions such as of urea groups with free NCO groups to biuret groups or of urethane units with free NCO groups to allophanate groups only take place in the absence of water and at higher temperatures. Under the applied test conditions it is assumed that a pure water reaction takes place, since the wood used for the bonds always provides a sufficient amount of water for the reaction due to its equilibrium moisture content (EMC) of about 12 %.

The cured polyurethane network is characterized by a segmented structure. The soft segments, consisting of flexible chains from the polyether oligomers, are joined by relatively rigid polyurethane-polyurea hard segments. The use of branched polyfunctional components (isocyanate and/or polyol components) ensures a high density of covalent (primary) cross-linking with higher thermal stability. A decomposition of urethane bonds, however, starts at about 120 to 250°C (Uhlig, 2006). Urea bonds have considerably higher thermal stability and start decomposing at about 260°C (Uhlig, 2006). The cross-linking of the hard segments is supported by hydrogen (secondary) bonding of the NH groups and carbonyl groups of urea and urethane linkages. Because of their stronger hydrogen bonding, polyurea hard segments tend to agglomerate into hard segment domains so that the structure of the

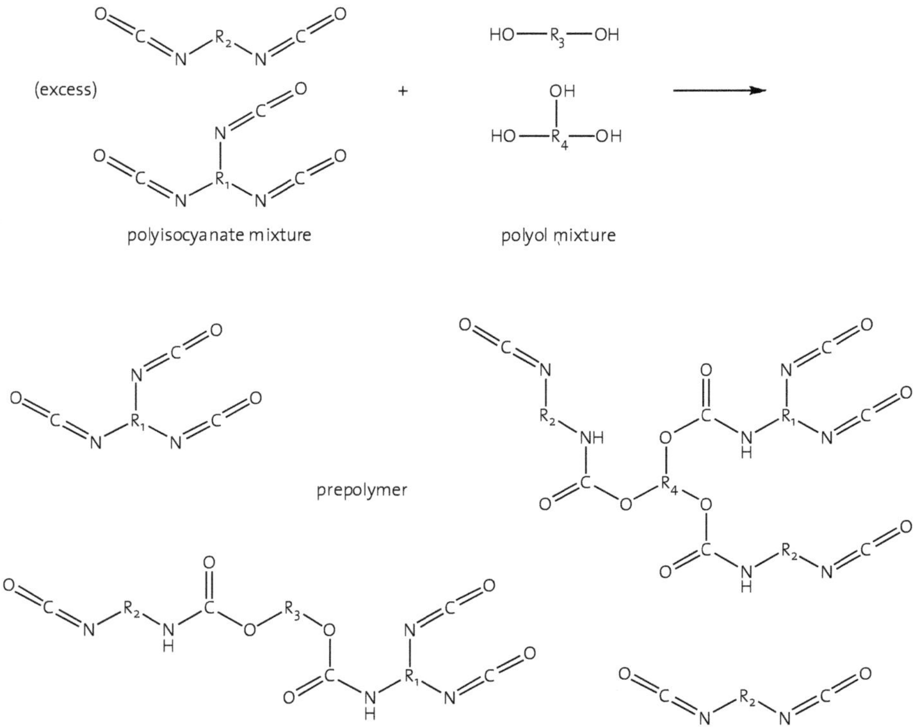

Fig. 3.2.3 Polyaddition reaction of isocyanate mixture and polyol mixture

polyurethane partly becomes separated into hard and soft segment-rich phases. Hydrogen bonds are also formed between NH groups of urea and urethane linkages with ether oxygens of the polyether chains but they are much weaker than those between hard segments. Secondary bondings exhibit a significantly lower bonding energy and thermal stability than primary bondings. In the polymer network, therefore, several structural bonding mechanisms interact with each other, thereby making it difficult to assign their individual contribution to a specific material behavior under temperature changes.

For this study, 20 laboratory prepolymers (hereinafter referred to as (A)-(W)) with varying chemical properties were produced by Bayer MaterialScience (Leverkusen, Germany). The parameters which are of particular interest are (i) the urea group content, (ii) the urethane group content and (iii) the cross-link density of the polyurethane network (Table 3.2.1). In addition, the hydrophilicity of the polyether mixture was stepwise increased by a higher (iv) EO content. Finally (v) the functionality

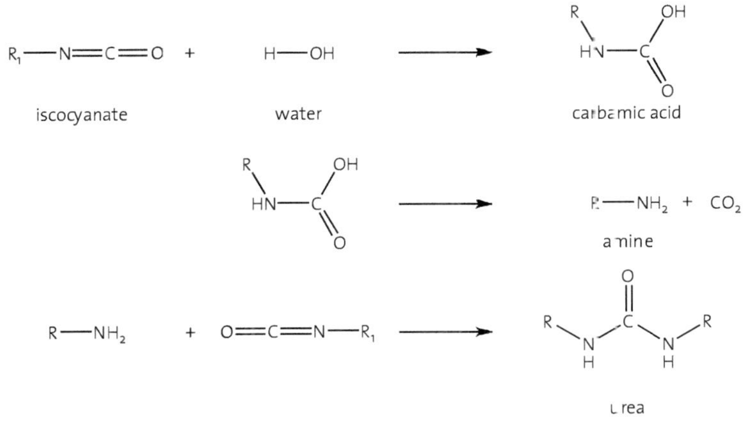

Fig. 3.2.4 Cross-linking of prepolymer with water to polyurethane

was adjusted, either by the isocyanate component or by the polyether component. To investigate the influence of the parameters mentioned, an experimental design was created where only one influencing parameter was changed at a time and the others remained constant, as far as possible taking the available raw materials into consideration.

i. The isocyanate content of the liquid prepolymer was varied in three steps to values of about 12, 16, and 20 % (the NCO content for adhesives commonly used in the wood industry is between 14 and 18 %). Since it is assumed that the NCO groups completely react with water to form urea groups, the urea group content can be easily assessed from the NCO content.

ii. The urethane group content was varied in three steps to values of about 0.23, 0.58 and 0.77 mol kg^{-1}. For each content two different configurations, using prepolymers with 16 and 20 % NCO, were adjusted. The EO content remained constant at 0 %.

iii. For each of the three urea group contents (1.48, 2.00 and 2.54 mol kg^{-1}), the cross-link density was also adjusted in three steps; however, taking the chemical composition into account it was not possible to reach equal values in the three groups of urea group content. The range aimed at was between 0.2 and 1.6 mol kg^{-1}, whereas the functionality was adjusted by the NCO component. The amount of urethane groups was 0.58 mol kg^{-1} for all mentioned prepolymers; the EO content was 0 %.

iv. The EO content was varied in three steps of about 0, 5 and 10 %. For each content two configurations with different urea group contents (2.00 and 2.54 mol kg^{-1}) were adjusted. The cross-link density remained nearly constant at between 0.21 and 0.27 mol kg^{-1}.

3 Main investigations

Table 3.2.1 Chemical prepolymer properties

Prepolymer	Experimental design[a]	NCO cont. [%]	PO cont. [%]	EO cont. [%]	NCO/PO/EO ratio [-]	Urea group cont.[b] [mol kg^{-1}]	Urethane group cont.[b] [mol kg^{-1}]	Cross-link density[b] [mol kg^{-1}]
A	i, iii, v	12	52.64	0	1/4.4/0	1.48	0.58	0.17
B	i, iii	12	51.59	0	1/4.3/0	1.48	0.58	0.64
C	i, iii	12	50.97	0	1/4.2/0	1.48	0.58	0.91
D	i, iii, iv, v	16	40.64	0	1/2.5/0	2.00	0.58	0.22
E	i, ii, iii	16	39.21	0	1/2.5/0	2.00	0.58	0.88
F	i, iii	16	38.40	0	1/2.4/0	2.00	0.58	1.25
G	i, iii, iv, v	20	28.82	0	1/1.4/0	2.54	0.58	0.27
H	i, iii	20	27.04	0	1/1.4/0	2.54	0.58	1.13
I	i, ii, iii	20	26.03	0	1/1.3/0	2.54	0.58	1.61
K	ii	16	45.29	0	1/2.8/0	2.00	0.23	0.87
L	ii	16	36.45	0	1/2.3/0	2.00	0.77	0.21
M	iv	16	32.78	4.83	1/2.0/0.3	2.00	0.77	0.21
N	iv	16	27.94	9.66	1/1.7/0.6	2.00	0.77	0.27
O	iv	20	21.01	4.79	1/1.1/0.24	2.54	0.77	0.27
P	iv	20	16.17	9.59	1/0.8/0.48	2.54	0.77	0.15
Q[c]	v	12	52.77	0	1/4.4/0	1.48	0.58	0.21
S[c]	v	16	40.66	0	1/2.5/0	2.00	0.58	0.27
U[c]	v	20	27.07	0	1/1.4/0	2.54	0.68	0.27
V	ii	20	30.97	0	1/1.5/0	2.54	0.23	1.62
W	ii	20	22.84	0	1/1.1/0	2.54	0.77	1.62

[a](i) variation of urea group content, (ii) variation of urethane group content, (iii) variation of cross-link density, (iv) variation of EO content, (v) variation of functionality adjustment
[b]Values apply for fully reacted polyurethane
[c]Functionality adjusted by the polyether component

Table 3.2.2 Wood properties

Species	Douglas fir	Spruce	Beech
Botanical name	*Pseudotsuga menziesii* F.	*Picea abies* Karst.	*Fagus sylvatica* L.
Category	Softwood	Softwood	Hardwood
Lignin content [%]	25 - 38	19 - 29	12 - 23
Cellulose content [%]	44 - 56	38 - 46	34 - 46
Oven-dry density [kg m^{-3}]	320 - 730	300 - 640	490 - 880
Raw density [kg m^{-3}]	350 - 750	330 - 680	540 - 910
Compression strength [MPa]	43 - 68	33 - 79	41 - 99
Bending strength[a] II [MPa]	68 - 89	49 - 136	74 - 210
Tensile strength[a] [MPa]	105	21 - 245	57 - 180
Shear strength (LT, LR) [MPa]	7.8 - 10.2	4.0 - 12.0	6.5 - 19.0
Young's modulus[a] II [GPa]	11.2 - 13.5	7.3 - 21.4	10.0 - 18.0

[a] in fiber direction

v. The functionality of the prepolymers was adjusted in two ways. For one group of three prepolymers with NCO contents of 12, 16 and 20 %, the functionality was adjusted by the NCO component; for another group with the same NCO contents, it was adjusted by the polyether component. The crosslink density remained nearly constant in the range from 0.17 to 0.27 mol kg^{-1}. The urethane group content was 0.58 mol kg^{-1}, except for prepolymer U which had about 0.68 mol kg^{-1}; the EO content for these prepolymers was 0 %.

Wood

The bondings were carried out with beech (*Fagus sylvatica* L.), spruce (*Picea abies* Karst.) and Douglas fir (*Pseudotsuga menziesii* Franco). Some important chemical and mechanical properties of these species are listed in Table 3.2.2. The softwoods spruce and Douglas fir, which are frequently used in engineered wood products, significantly differ from hardwoods such as beech regarding their anatomical structure as well as in their mechanical properties. Because of its higher strength and an extractive content of about 2 %, beech is well suited for adhesive testing. In the case of too low shear strength, the wood strength and not the strength of the adhesive would be evaluated. That is the reason why beech is required by standards like DIN EN 302-1 (2004).

Preparation of specimens

The tensile shear strength was determined according to DIN EN 302-1 (2004). Therefore, lap joints were produced from prefabricated boards that were previously stored under standard climatic conditions (20°C, 65 % RH) until the EMC was reached. To avoid aging effects on the wood surface, the boards were planed to the necessary thickness of (5 ± 0.1) mm immediately before the bonding process. The bonding was performed with close-contact bond lines ($\approx$0.1 mm), with the adhesives therefore being applied on one side with a toothed spatula and a spread of $200\,\mathrm{g\,m^{-2}}$. The pressing was conducted at room temperature and 50 % RH at a bonding pressure of about 0.8 MPa. Due to the different open times of the prepolymers, all bondings were conducted with a pressing time of 12 h. After one week of storage under standard climatic conditions, the bonded adherends were cut into specimens according to the standard.

Prepolymer films were produced by applying the liquid adhesive to a PE sheet. During the reaction with the humidity in the surrounding air, carbon dioxide (CO_2) is released and forms gas bubbles within the film. Therefore, it is necessary to consider the maximum permissible film thickness of about 0.2 mm and to ensure a reaction under optimal climatic conditions. In our case, 20°C and 50 % RH were most suitable. After peeling-off from the PE sheet, the films were stored under standard climatic conditions for a minimum of three days.

Tensile shear tests on bonded wood joints

The shear tests were performed position-controlled until failure under standard climatic conditions, using a Zwick Z100 universal testing machine. After the test, the wood failure percentage (WFP) was estimated visually in 10 % steps. The mean tensile shear strength (τ) was calculated from 15 repititions of each combination.

Preliminary tests were performed with spruce, Douglas fir and beech. After analyzing these results, it was appropriate to investigate the temperature influence on beech only. The temperature during the test was varied in four steps (40, 70, 150 and 200°C). In addition, the tempering time was subdivided into two durations of 2 and 4 h. The tests under increased temperatures took place immediately after each single sample was taken out of the oven.

Tensile tests on polymer films

The determination of the tensile properties of the adhesives was carried out according to DIN EN ISO 527-1 (1996). The actual samples were punched using sample shape type 1B specified in DIN EN ISO 527-3 (2003). A Zwick/Roell Z100 testing machine with a 500 N load cell and a test speed of $5\,\mathrm{mm\,min^{-1}}$ was used to perform a displacement-controlled test under standard climatic conditions. The strain was measured optically by means of video-extensometry. Besides Young's modulus, tensile strength and maximum strain were evaluated.

Dynamic mechanical analysis (DMA)

Samples of 20 mm length and 4 mm width were cut from the prepared adhesive films and evaluated in a Seiko DMS 210 apparatus in the tensile mode over a temperature range of -140 to 250°C, at a frequency of 1 Hz, and at a ramp rate of 2°C min^{-1}. The underlying standard for this test is ISO 6721-4 (1994).

Differential scanning calorimetry (DSC)

The thermal properties of the polyurethane prepolymer films were analyzed using a Perkin-Elmer DSC-7 differential scanning calorimeter. Approximately 10 mg of polyurethane films were placed in standardized pans with caps using a heating rate of 20°C min^{-1}. Two consecutive runs were carried out, heating from -100 to 100°C followed by cooling down to -100°C (cooling rate 320°C min^{-1}) and nitrogen flushing before the second heating run from -100 to 100°C. The glass transition temperatures (T_g) were determined at half height of the glass step.

Viscometry

The viscosity of the prepolymers was determined at 23°C with a MCR 301 cone/plate rheometer (d=25 mm, $\alpha = 1\check{r}$) at a shear rate of $150\,s^{-1}$ according to DIN 53019 (2008).

Film formation

The film formation times and film drying times of the prepolymers were recorded using a BK drying time recorder at 23°C and 50 % RH.

3.2.4 Results and discussion

Wood properties

Beech wood achieved a considerably higher tensile shear strength than the softwoods spruce and Douglas fir (Fig. 3.2.5). The mean values (beech: 16.20 MPa, spruce: 9.77 MPa, Douglas fir: 11.44 MPa) at standard climatic conditions were higher than values presented by Wagenführ (2004), which is most likely why the values were not estimated by the standard for plain wood (DIN 52187, 1979). The high variance of tensile shear strength (coefficient of variation up to 25 %) of the species was caused by a selection of different raw materials and by the band width of annual ring angles (from 30° to 90°), which, especially in the case of beech wood, yields different values between the LR and LT planes (Haß et al., 2009).

The main focus was put on the temperature-dependent shear strength. The graphs display a decrease of τ in the temperature range from 20 to 200°C. Only Douglas fir showed a lower value at 20°C compared to 40°C. However, the difference is not statistically significant. As under standard climatic conditions, the softwoods spruce and Douglas fir achieved lower values compared to beech also at elevated temperatures. The decrease of τ up to a maximum temperature of 200°C amounted to 48 %

3 Main investigations

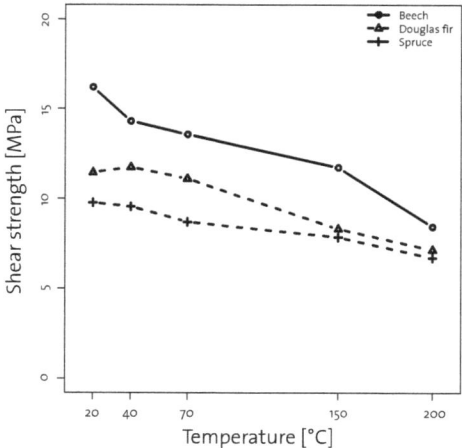

Fig. 3.2.5 Tensile shear strength of solid wood vs. temperature

of the maximum strength for beech, 32 % for spruce and 38 % for Douglas fir. Besides the significant main effects temperature, species and time, the analysis of variance (ANOVA) showed a significant interaction between species and temperature. Due to the different structure of softwoods and hardwoods, and due to the different contents of the main components lignin, cellulose and hemicelluloses, differences in the curve progressions are not surprising. Besides temperature, the wood moisture also has a strong influence on strength development (Kudela, 1996). Both effects are hardly separable experimentally since - due to the tempering of wood - a part of the included water is always released. At a temperature of about 150°C, the specimens were completely oven-dry. Kudela (1996) investigated this moisture-temperature interaction for beech wood and quantified the influences. Between 2 and 4 h of tempering, no statistically significant difference regarding the tensile shear strength was obtained by ANOVA ($p<0.05$).

Prepolymer properties

Urea group content The variation of the urea group content showed distinct differences in tensile shear strength of bonded joints as well as in the storage modulus of the prepolymer films. At 70, 150 and 200°C, a higher urea group content caused significantly higher shear strength (Fig. 3.2.6(a)). The difference between low and medium urea group content was more pronounced than between medium and high urea group content. However, higher values of τ are hardly attainable as this is limited by the maximum wood strength reached at this level. The curves of low and medium urea group content showed tendentially up to 40°C an increase of τ; at high urea group content the increase went up to 70°C; considering the long open times of the unformulated adhesives, this may be attributed to an aftercure as a consequence of water release during heating up. Even though the EMC of wood amounted to 12 %, the water is not unlimitedly available for the reaction due to chemical and/or physical bonds to the wood structure. After reaching the maximum, the tensile shear strength decreased nearly linearly in the same way as solid wood. The statistical analysis documented a highly significant influence of the factors urea group content and temperature. The exposure time in the oven (2 or 4 h) had no significant effect in this test; therefore the results of both groups were combined in the plots.

A comparison of the storage modulus curves (Fig. 3.2.6(b)) of the films clearly reveals differences between the three groups of prepolymers regarding their temperature dependency. The prepolymers with the lowest urea group content dropped already in the negative temperature range at about -50°C and decreased almost linearly up to a temperature of 200°C. By means of DSC in the range of -50°C a T_g was found for these prepolymers (Table 3.2.3) which characterizes the glass transition of the soft segments in the polymer structure. Besides the decrease in the negative temperature range, the storage modulus of all prepolymers showed a second shoulder in the storage modulus curves at about 50°C. This step again seems to reveal a T_g of the polymer, caused by the dissociation of hydrogen bonds. The hydrogen bonds of the urethane groups dissociate earlier due to the lower electrostatic attraction of the oxygen atom compared to the nitrogen atom of the urea group. As reported in (Daniel da Silva, Martín-Martínez and Bordado, 2006) an increase in the free NCO content results in decreased values of T_g due to the reduction in the average molecular weight. Between 45 and 85°C, the DSC results showed a broader T_g range compared to the dynamic method. The differences can be ascribed to the different methods since the DMA frequency and the DSC heat rate influence the results. Above this glass step, the hard segments appeared to be solved in the soft segment matrix.

The results of the tensile test (Table 3.2.3) support the measurements made by DMA and tensile shear test. At 20°C, the Young's modulus and maximum strength of films increased tendentially with increasing urea group content. An ANOVA confirms the observation; the effect of urea group content on the Young's modulus was found to be significant on the 5 % level of significance. For all other predictor variables no significant effects were found. The fracture behavior of the films revealed for low urea group contents generally ductile fracture without showing a yield point. For high urea group content on the other hand, generally brittle fracture occurred.

The only prepolymers that revealed a significant WFP at 200°C were the types H and I with high urea group content and also high cross-link density. In the case of these prepolymer configurations it was possible to overcome the wood cohesion strength.

An increase of the NCO content caused lower viscosity (Table 3.2.3) of the prepolymers due to the decreased average chain length and less intermolecular interactions between polymer chains (Daniel da Silva, Martín-Martínez and Bordado, 2006). Negative effects on the tensile shear strength due to possible higher penetration of the low viscous prepolymers could not be detected.

3 Main investigations

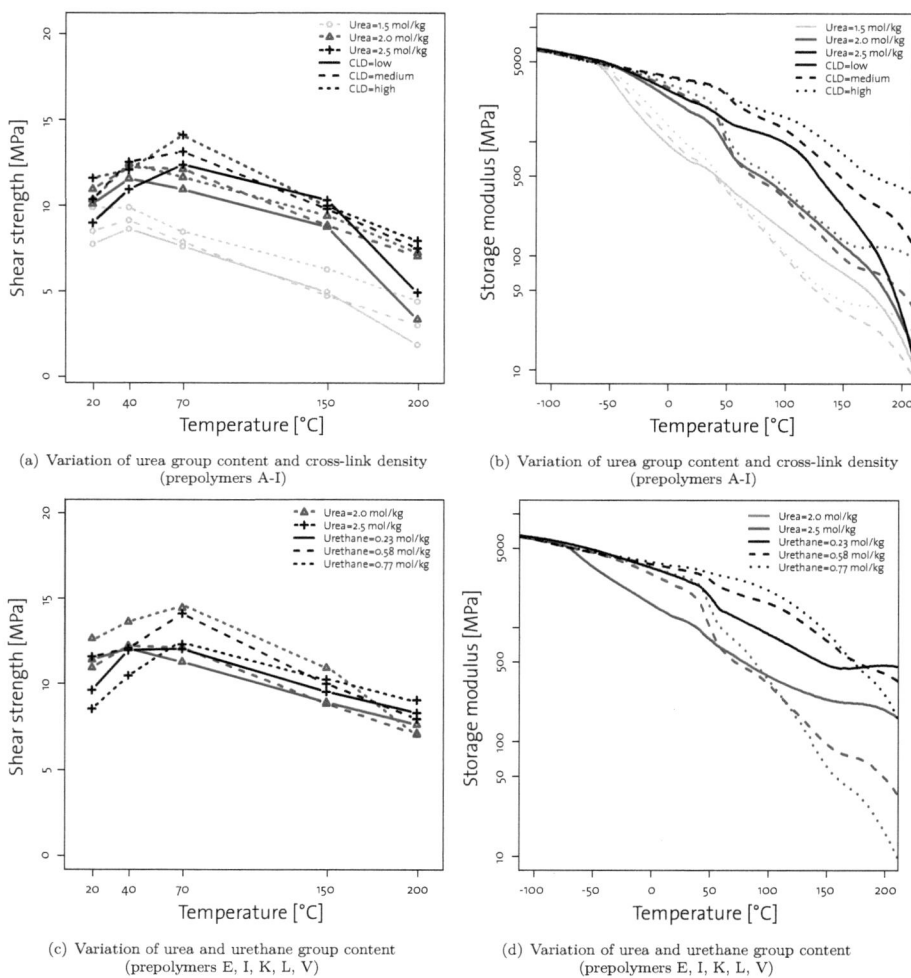

Fig. 3.2.6 Tensile shear strength of bonded wood joints and storage modulus of films vs. temperature

Cross-link density The results showed that the cross-link density had a minor influence on the tensile shear strength of bonded wood joints (Fig. 3.2.6(a)). The ANOVA, however, indicated a statistically significant effect for this factor and also for the interaction of temperature and cross-link

Table 3.2.3 Film formation properties of liquid prepolymers and results of tensile test and DSC on prepolymer films

Prepolymer	T_{g1} [°C]	T_{g2} [°C]	μ (23°C) [mPas]	t_f [min]	t_d [min]	E [MPa]	σ_{max} [MPa]	ε [%]
A	-42.9	-	2610	360	1140	130.0	32.6	309.9
B	-23.4	51.9	5290	480	720	489.3	25.9	64.4
C	-27.2	52.2	11100	420	570	396.4	32.8	98.6
D	-	47.5	1310	300	930	989.7	40.3	174.0
E	-	53.1	3690	420	720	2309.6	65.0	13.7
F	-	54.5	5460	420	630	3518.2	104.4	16.4
G	-	57.2	617	360	930	-	-	-
H	-	60.3	2540	390	730	2595.2	75.0	4.2
I	-	80.9	4810	450	720	2513.5	71.9	17.3
K	-59.7	-	1620	360	630	784.6	37.0	75.3
L	-	-	8920	420	720	2085.3	58.9	7.6
M	-	-	2130	540	840	2439.8	69.5	8.3
N	-	-	2032	510	720	1899.0	55.0	4.2
O	-	74.3	1020	270	840	-	-	-
P	-	-	915	360	750	-	-	-
Q	-49.3	-	3450	270	960	203.8	24.4	226.8
S	-38.4	56.5	1810	330	900	727.0	26.0	80.8
U	-	60.4	1050	270	690	-	-	-
V	-	58.5	2450	480	780	2014.2	63.4	7.7
W	-	62.3	12000	480	720	2192.0	56.9	4.1

T_g glass transition temperature, μ viscosity, t_f film formation time, t_d film drying time, E Young's modulus, σ_{max} tensile strength, ε_{max} maximum strain

density at a 5 % level of significance. Only at 200°C did a differentiation between the prepolymers with low and those with medium or high cross-link density respectively become possible.

The addition of polymeric MDI to increase the cross-link density had a large effect on the viscosity that increased significantly with increasing functionality (Table 3.2.3). A higher functionality further accelerated the film formation due to faster molar mass development. The bonding quality was not influenced by the film formation properties. In the case of storage modulus (Fig. 3.2.6(b)), a tendency similar to that of the tensile shear strength was revealed. Only at temperatures above 150°C did the better performance of prepolymers with higher cross-link density become evident. The differences were more pronounced for the prepolymers with a higher urea content. Obviously the cross-link density had no influence on the modulus at high temperatures as long as the urea content was below a certain threshold.

By means of DSC measurements, an increase in T_g with increasing cross-link density was found. At higher temperatures the modulus was mostly dominated by the covalent network bonds, and hydrogen bonds had no influence on stiffness.

3 Main investigations

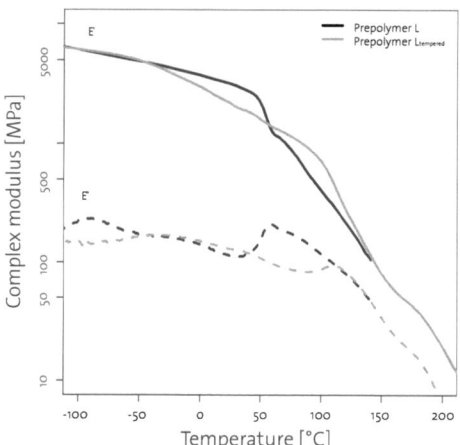

Fig. 3.2.7 Storage and loss modulus of prepolymer L in normal state and tempered at 100°C for 1 h

Urethane group content The results of the tensile shear strength revealed that the urethane group content is statistically significant but the measured effect was less pronounced. According to the graphs in Fig. 3.2.6(c) almost no difference between the groups of prepolymers with varying urethane group content can be detected. Although significantly higher values for the prepolymer L were found in the temperature range from 20 to 150°C compared to the remaining prepolymers, no lucid explanation could be found for this effect. All prepolymers again showed an increase of tensile shear strength in the lower temperature range up to 40 or 70°C, respectively. With further increasing temperatures, the shear strength decreased linearly.

The DMA graphs in contrast (Fig. 3.2.6(d)) show, besides the influence of the urea group content, a distinct influence of the urethane group content. In the case of prepolymer K, with the lowest urea and urethane group content, the dominance of the soft segment structure is shown by the decrease of storage modulus at about -60°C which reveals the T_g of the polyether component. All other prepolymers revealed a T_g at around 50°C. Similar T_g as for DMA were revealed in the first heating in DSC measurements (Table 3.2.3).

Surprisingly, the second heating in DSC measurements did not show a glass transition temperature, although DMA measurements after annealing at 100°C showed a broad glass transition with a maximum in loss modulus at about 110°C (Fig. 3.2.7). This behavior seems to be similar to the plasticizing effect of water in polyamide. Polyamide 6 stored at 50 % relative humidity has a glass transition of about 0°C, whereas dried polyamide 6 has a glass transition temperature of about 50°C (Illers, 1960).

The polar bonds, developed between NH groups and carbonyl groups of urethane linkages, cause strong physical cross-linking which is why also the viscosity of the prepolymers depends primarily on the urethane group content. Up to a certain temperature, the prepolymers with higher urethane group content also revealed higher storage moduli. Since the higher urethane group content resulted from using ether components with lower molecular weight, the urethane groups can move closer together and therefore the stiffness of the network was enhanced. However, above 100°C a higher urethane group content had negative effects on the mechanical properties for prepolymers with a urea group content of $2\,\mathrm{mol\,kg^{-1}}$. The prepolymers with a urea group content of about $2.54\,\mathrm{mol\,kg^{-1}}$ revealed this turnaround at about 180°C. This could be caused by urethane groups that start to dissociate in the temperature range of about 150°C.

EO content   The graphs in Fig. 3.2.8(a) show that a differentiation between the prepolymers regarding tensile shear strength by means of the EO content was not possible. The measurement of the storage modulus, however, revealed distinct differences in the temperature-dependent propagation (Fig. 3.2.8(b)). All prepolymers displayed a T_g at about 50°C. The lower the EO content, the less pronounced the decrease of storage modulus with increasing temperature. According to that, prepolymers without EO ethers in the mixture had a better thermal stability. It is probably a restricted mobility of the PO ether due to the methyl moiety in the chain that caused this effect. Because of the low cross-link density of the prepolymers in this experimental design, the stiffness of the films as well as the tensile shear strength were below average.

Adjustment of functionality   Fig. 3.2.8(c) shows the mean tensile shear strength vs. temperature of three groups of prepolymers with different urea group content but the same adjustment in each group. Regarding thermal stability it is not relevant whether the functionality of the prepolymer was adjusted by the isocyanate or the polyether component. Both types of prepolymers showed nearly the same temperature-dependent behavior. After reaching the maximum in the range between 40 and 70°C, the shear strength decreased almost linearly with increasing temperature. Only the prepolymers with the lowest urea group content revealed by trend better shear strengths for the adjustment via polyether. A possible reason for a better performance could be an increase of stiffness due to the cross-link points of the ether component which stabilize the soft segment.

The storage modulus, however, revealed - for low and medium urea group content - a lower stiffness for the polyether adjustment than for the NCO adjustment (Fig. 3.2.8(d)). The propagation of the different curves is in principle similar.

Statistical analysis

A multiple linear regression was performed to analyze the relationship between the response variable tensile shear strength and the predictor variables. Furthermore the statistical effects and their effect sizes were determined. The multiple linear model for relationships between response and predictor variables, accounting only for main effects, took the form: $Y = \beta_0 + \beta_1 x_1 + \beta_2 x_2 + \beta_3 (x_3)^2 + \beta_3 x_3 +$

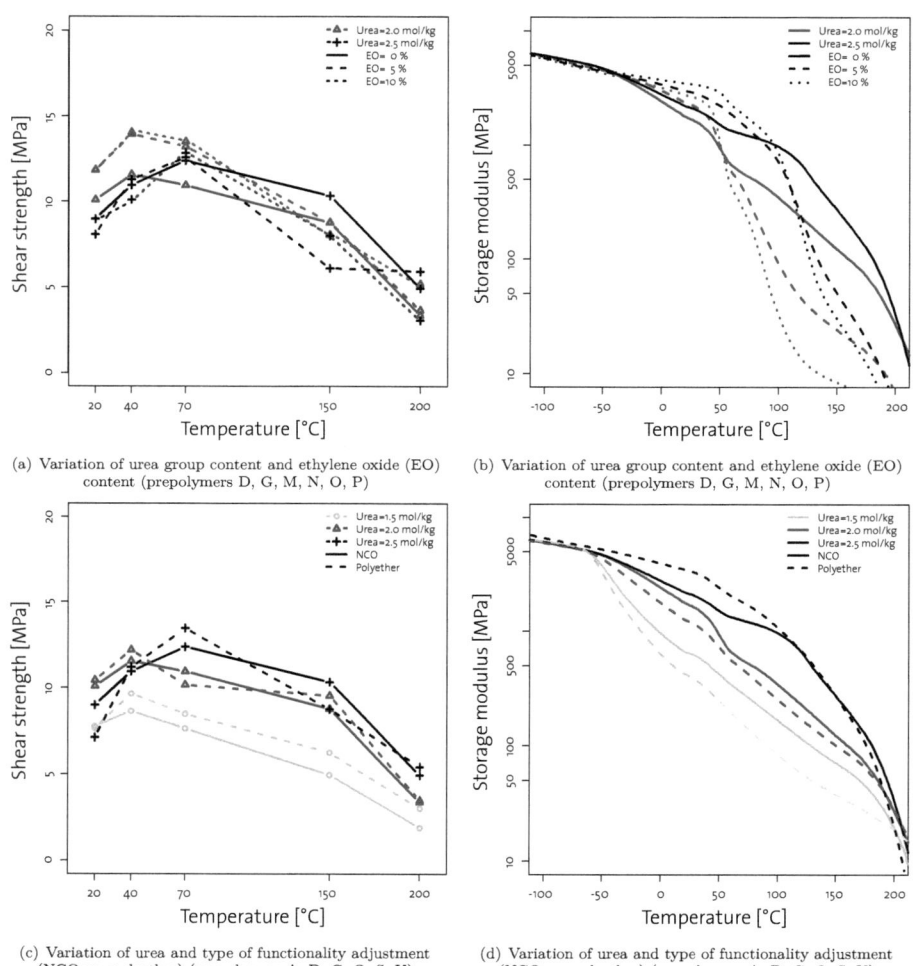

Fig. 3.2.8 Tensile shear strength of bonded wood joints and storage modulus of films vs. temperature

$\beta_4 x_4 + \beta_5 x_5 + \beta_6 x_6 + \varepsilon$, where Y is the response variable tensile shear strength, $x_1, \ldots, x_6$ the predictor variables (x_1 cross-link density, x_2 urea group content, x_3 temperature, x_4 time, x_5 urethan group content, x_6 EO content), β_0 the intercept and ε the error term. Uncontrollable variables such as wood density, growth ring angle and growth ring width were not taken into account since preferably adhesive

Table 3.2.4 Coefficients of regression

	Estimate	Std. error	t-value	$Pr(>|t|)$	
(Intercept)	5.14e+00	3.96e-01	12.983	< 2e-16	***
Cross-link density	1.18e+00	1.06e-01	11.172	< 2e-16	***
Urea group content	2.23e+00	1.28e-01	17.501	< 2e-16	***
Temperature^2	-2.18e-04	1.85e-05	-11.776	< 2e-16	***
Temperature	1.41e-02	4.45e-03	3.173	0.00153	**
Time	4.56e-02	4.46e-02	1.022	0.30705	
Urethane group content	9.74e-01	3.50e-01	2.786	0.00538	**
EO content	3.38e-02	1.84e-02	1.833	0.06699	.

Residual standard error: 2.12 on 2262 degrees of freedom, multiple R^2: 0.62, adjusted R^2: 0.62, F-statistic: 522.4 on 7 and 2262 DF, p-value: < 2.2e-16, signif. codes: 0 '***' 0.001 '**' 0.01 '*' 0.05 '.' 0.1 ' ' 1

failure appeared. Since the correlation between response variable and temperature is non-linear, beside the linear term also a squared term was included in the model.

Table 3.2.4 shows the summary of the statistical analysis. The main effects cross-link density, urea group content, temperature, temperature squared and urethane group content were found to be significant. The predictor variables time and EO content on the other hand did not influence the tensile shear strength significantly. The adjusted R^2 for the model amounted to 0.62, thus the model accounts for 62 % of the variance.

A quantification of the effects (Table 3.2.5) revealed the temperature to have the biggest effect on the tensile shear strength. The increase in temperature from 20 to 200°C caused a decrease in shear strength of 68 %, under the condition that all other predictor variables remained constant at the lowest value. Regarding the chemical parameters, the urea group content had the strongest effect on the tensile shear strength. An increase from 1.48 to 2.54 mol kg^{-1} caused an increase in tensile shear strength of about 26 %. The cross-link density was the second most influencing parameter with an increase of 19 % from the lowest to the highest value. The urethane content was responsible for an increase of about 7 %. If the temperature increases to 200°C, the effects of the chemical parameters are more than twice as high. This fact suggests interactions between the chemical parameters and the temperature. If these interactions are considered in the model, all of them were found to be significant; however, the coefficient of determination is increased only by 1 %.

The residuals were analyzed for the response variable tensile shear strength. The quantile-quantile plot (Fig. 3.2.9(a)) shows that the residuals for the model were normally distributed. The residual dispersions for the model were randomly distributed, and the residuals were approximately evenly split and close to the average value of the response variable (Fig. 3.2.9(b)). The absence of a systematic pattern in the residuals confirms that the model found is adequate with respect to the response variable.

3 Main investigations

Table 3.2.5 Quantification of the statistical effects by prediction of tensile shear strength

Variable	$x_{i,min}$	$x_{i,max}$	t_{min} [MPa]	t_{max} [MPa]	$\Delta\tau$ [MPa]	$\Delta\tau$ [%]
Cross-link density [mol kg^{-1}]	0.15	1.62	9.17	10.91	1.73	18.91
Urea groups [mol kg^{-1}]	1.5	2.5	9.17	11.4	2.23	24.31
Temperature [°C]	20	200	9.17	3.08	-6.09	-66.45
Time [h]	2	4	9.17	9.26	0.09	0.99
Urethane groups [mol kg^{-1}]	0.23	0.77	9.17	9.70	0.53	5.73
EO content [%]	0	9.66	9.17	9.50	0.33	3.56

(x_i) domain of respective variable, (τ_{min}) minimum estimated shear strength, (τ_{max}) maximum estimated shear strength, $\Delta\tau$ difference between τ_{min} and τ_{max}

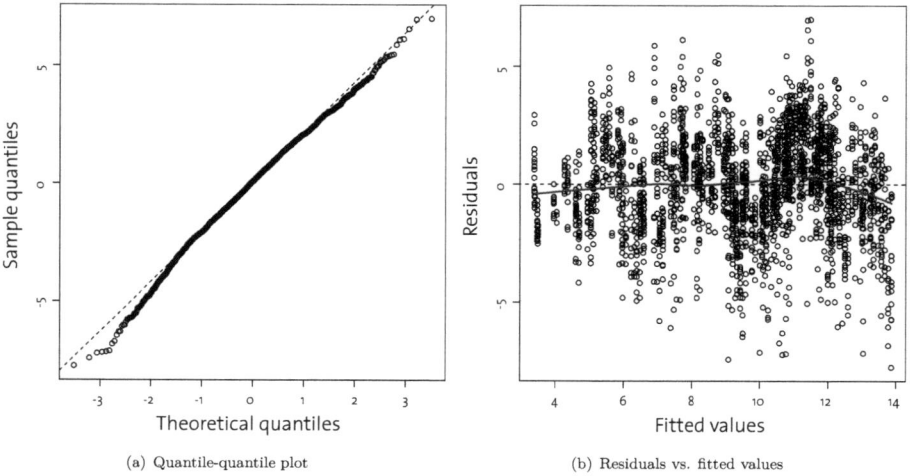

(a) Quantile-quantile plot   (b) Residuals vs. fitted values

Fig. 3.2.9 Residual analysis

3.2.5 Conclusions

The thermal stability of moisture-curing 1C PUR prepolymers is mainly influenced by the urea group content and the cross-link density of the reacted system. The shear strength of bonded wood joints as well as the stiffness of prepolymer films is significantly increased by a higher content of urea hard segments, nearly independent of the temperature. The higher the content of hard segments in the polymer structure the more the polymer properties are dominated by this domain. The increase of urethane hard segments, on the other hand, has only a minor effect on the measured properties. The results suggest that at temperatures below 100°C higher urethane content contributes positively to the tensile shear strength of bonded joints as well as to the storage modulus of films for the same reason as the urea content. However, at temperatures above 180°C, the increased urethane content has rather negative effects; the reason may lie in the primary degradation of the urethane groups.

An improvement in the thermal stability by using EO ethers is not expected. On the contrary, ether mixtures consisting only of PO ethers reveal higher stiffness after passing the glass step in the range 50 to 100°C.

Both types of functionality adjustment (by polyether or isocyanate component) are common in practice, depending on criteria such as color, raw material availability and price, and by means of the results there is no reason to prefer one of the types regarding thermal stability. The analyzed prepolymers showed an increase in tensile shear strength from ambient conditions up to temperatures of 70°C, which is most likely caused by tempering effects. Due to insufficient sorbate diffusion into the glue line at ambient conditions, not all of the free NCO groups react with water. Higher temperatures may provide higher molecular movement and account for a complete reaction.

In principle, the properties of the prepolymers can largely be modified by varying the molar ratio of the components. The polyisocyanate component is mostly responsible for the properties (elasticity, cohesion and strength) of the finished product. Further modification possibilities are given by the subsequent formulation (catalyst, filler, plasticizer, etc.), which will be discussed in a following paper (Clauß et al., 2011).

The processing conditions of the adhesive are influenced by the prepolymer configuration as well. Therefore it is necessary to find a compromise between the mechanical properties of the finished product and the application properties such as viscosity and reactivity. Compared to other types of adhesives, NCO-terminated prepolymers offer advantages regarding formulation, such as longer pot-life in the absence of water, increased initial bond strength, almost no low-molecular substances in the cured product and a more favorable balance between the components in the mixture.

3.2.6 Acknowledgments

The authors acknowledge the financial support of the Fund for the Promotion of Forest and Wood Research. Special thanks are directed to Dr. E. Mayer for his valuable contribution to this research.

References

ASTM D 7247 (2007) Standard test method for evaluating the shear strength of adhesive bonds in laminated wood products at elevated temperatures. West Conshohocken, PA: ASTM International

Chattopadhyay, D.K. and Webster, D.C. (2009) Thermal stability and flame retardancy of polyurethanes. Prog. Polym. Sci. 34 No. 10, 1068–1133

Clauß, S., Allenspach, K. and Niemz, P. and Gabriel, J. (2011) Improving the thermal stability of one-component polyurethane adhesives by adding filler material. Wood Sci. Technol. 45 No. 2, 383–388

Clauß, S. et al. (2011) Influence of the adhesive formulation on the mechanical properties and bonding performance of polyurethane prepolymers. Holzforschung, 65 No. 6, 835–844

Clauß, S., Joščák, M. and Niemz, P. (2011) Thermal stability of glued wood joints measured by shear tests. Eur. J. Wood Prod. 69 No. 1, 101–111

Daniel da Silva, A.L., Martín-Martínez, J.M. and Bordado, J.C.M. (2006) Influence of the free isocyanate content in the adhesive properties of reactive trifunctional polyether urethane quasi-prepolymers. Int. J. Adhes. Adhes. 26 No. 5, 355–362

DIN 52187 (1979) Testing of wood – determination of ultimate shearing stress parallel to grain. Berlin: Beuth Verlag

DIN 53019 (2008) Viscometry – Measurement of viscosities and flow curves by means of rotational viscometers. Berlin: Beuth Verlag

DIN EN 14257 (2006) Adhesives – Wood adhesives – Determination of tensile strength of lap joints at elevated temperature (WATT'91). Berlin: Beuth Verlag

DIN EN 14292 (2005) Adhesives – Wood adhesives – Determination of static load resistance with increasing temperature. Berlin: Beuth Verlag

DIN EN 302-1 (2004) Adhesives for load-bearing timber structures – Test methods – Part 1: Determination of bond strength in longitudinal tensile shear strength. Berlin: Beuth Verlag

DIN EN ISO 527-1 (1996) Plastics – Determination of tensile properties – Part 1: General principles. Berlin: Beuth Verlag

DIN EN ISO 527-3 (2003) Plastics – Determination of tensile properties – Part 3: Test conditions for films and sheets. Berlin: Beuth Verlag

Frangi, A., Fontana, A. and Mischler, A. (2004) Shear behaviour of bond lines in glued laminated timber beams at high temperatures. Wood Sci. Technol. 38 No. 2, 119–126

Frangi, A. and Fontana, M. (2003) Charring rates and temperature profiles of wood sections. Fire Mater. 27 No. 2, 91–102

George, B. et al. (2003) Comparative creep characteristics of structural glulam wood adhesives. Holz Roh Werkst. 61 No. 1, 79–80

Haß, P. et al. (2009) Influence of growth ring angle, adhesive system and viscosity on the shear strength of adhesive bonds. Wood Mater. Sci. Eng. 4 No. 3, 140–146

Illers, K.-H. (1960) Der Einfluß von Wasser auf die molekularen Beweglichkeiten von Polyamiden. Makromol. Chem. 38 No. 1, 168–188

ISO 6721-4 (1994) Plastics – Determination of dynamic mechanical properties – Part 4: Tensile vibration - Non-resonance method. Berlin: Beuth Verlag

Kudela, J. (1996) Influence of moisture and temperature loading on strength of beech wood loaded in compression. Drev. Vysk. 41 No. 2, 3–17

Na, B. et al. (2005) One-component polyurethane adhesives for green wood gluing: Structure and temperature-dependent creep. J. Appl. Polym. Sci. 96 No. 4, 1231–1243

Niemz, P. and Allenspach, K. (2009) Studies on the influence of temperature and timber moisture on the failure behaviour of selected adhesives under tensile shear load. Bauphysik, 31 No. 5, 296–304

Properzi, M. et al. (2002) Comparative performance of fast-setting single and separative application exterior wood adhesives for stuctureal glulam. Holzforschung und Holzverwertung, 54 No. 1, 18–20

Richter, K et al. (2006) Thermal Stability of structural one-component polyurethane adhesives. In **Friehard, C.R., editor:** Wood Adhesives 2005. Madison, WI: Forest Products Society, 203–210

Richter, K., Pizzi, A. and Despres, A. (2006) Thermal stability of structural one-component polyurethane adhesives for wood - Structure-property relationship. J. Appl. Polym. Sci. 102 No. 1, 24–32

Richter, K. and Steiger, R. (2005) Thermal stability of wood-wood and wood-FRP bonding with polyurethane and epoxy adhesives. Adv. Eng. Mater. 7 No. 5, 419–423

Schrödter, A. and Niemz, P. (2006) Investigation on the failure behaviour of glue joints at high temperatures and relative humidity. Holztechnologie, 47 No. 1, 24–32

Šebenik, U. and Krajnc, M. (2007) Influence of the soft segment length and content on the synthesis and properties of isocyanate-terminated urethane prepolymers. Int. J. Adhes. Adhes. 27 No. 7, 527–535

Sonderegger, W., Hering, S. and Niemz, P. (2011) Thermal behaviour of Norway spruce and European beech in the principal anatomical directions. Holzforschung. 65 No. 3, 369–375

Uhlig, K. (2006) Polyurethan Taschenbuch. 3rd edition. München: Hanser, 208 p.

Wagenführ, R. (2004) Holzatlas. 2nd edition. München: Hanser, 371 p.

Woods, G. (1990) The ICI polyurethanes book. New York: John Wiley, 362 p.

Yeh, B et al. (2006) Adhesive performance at elevated temperatures for engineered wood products. In **Frihart, C.R., editor:** Wood Adhesives 2005. Madison, WI: Forest Products Society, 195–201

3.3 Paper III

Holzforschung (2011) 65:835-844

Influence of the adhesive formulation on the mechanical properties and bonding performance of polyurethane prepolymers

Sebastian Clauß[1], Joseph Gabriel[2], Alexander Karbach[3], Mathias Matner[4], Peter Niemz[1]

[1]Institute for Building Materials, ETH Zurich, Schafmattstrasse 6, 8093 Zurich, Switzerland
[2]Purbond AG, Industriestrasse 17a, 6203 Sempach-Station, Switzerland
[3]CURRENTA GmbH & Co. OHG CHEMPARK, Solid State and Polymer Analysis, Rheinuferstrasse 7-9, 47829 Krefeld, Germany
[4]Bayer MaterialScience AG, Coatings, Adhesives & Specialties, Kaiser-Wilhelm-Allee 1, 51373 Leverkusen, Germany

3.3.1 Abstract

Only small amounts of additives are needed to formulate one-component polyurethane (1C PUR) adhesives for various applications. The current study illuminates the effects of the formulation on the mechanical properties of pure adhesives, on the one hand, and their performance in bonded wood joints on the other. Tensile shear tests on bonded wood joints, tensile tests on adhesive films, and nanoindentation measurements in the interphase region of the bond were performed. Analyses by means of infrared, atomic force, and electron microscopy provided the explanatory basis for the results obtained. Additionally to laboratory made 1C PUR, unmodified commercial 1C PUR, melamine-urea-formaldehyde (MUF), and phenol-resorcinol-formaldehyde (PRF) were tested for comparison. The results obtained confirm that the mechanical properties of 1C PUR adhesives are significantly affected by their prepolymer composition. The adhesive formulation by means of additives, on the other hand, does not affect the mechanical properties but is to a large extent responsible for the bonding performance.

3.3.2 Introduction

Adhesives for modern timber engineering should have a broad range of properties, fitting to variations of wood species, process conditions, and application requirements. The rheological and kinetic properties of the adhesive have to be adjusted to the anatomical features of wood as an anisotropic biocomposite. The penetration in longitudinal direction, for example, is usually higher than that perpendicular to the grain (Siau, 1984). Softwood and hardwood properties vary considerably; significant variations also exist between single species (Kamke and Lee, 2007). Species with a higher density reveal lower penetration due to a smaller lumen size, which results in limited interlocking of the adhesive. A higher density also causes greater dimensional changes (swelling and shrinking), which are leading to higher internal stresses (Suomi-Lindberg, Pulkkinen and Nussbaum, 2002). In addition, wood extractives may reduce the wettability of the surface as well as the flow and penetration of the adhesive (Hse and Kuo, 1988). Mechanical wood properties vary also considerably between species and change under environmental influences, such as moisture and temperature (Niemz, 1993).

The fabrication of finger joints, straight and bent glulam beams, cross-laminated timber, or of other structural wood products needs adhesives that are tailor-made in terms of their open time, pressing time, and rheology. The conditions in which the wooden joints are used also differ. Adhesives for load-bearing timber constructions, for example, must generally resist high static and dynamic mechanical loads, as well as high stresses due to the swelling or shrinking of wood resulting in increased elastic and even plastic deformations.

Polyurethane-based adhesives (PUR) and sealants have been known for a long time as versatile tools in wood industry. 1C PUR adhesives have good strength, are ductile, resistant to moisture, and cause little emission. The cross-linking reaction of the isocyanate groups (-NCO) occurs under the influence of water derived from the moisture content (MC) of the wood or from the humidity in the ambient air. This reaction leads under CO_2 formation to thermally stable ureatype structural groups, which build up strong hard segments. Under dry conditions, high temperatures, influence of filling material, various undesired structures – such as allophanates, biurets, uretdiones or carbodiimides – can be formed (Ionescu, 2005). These reactions may influence the cross-link density of the polyurethane network (Woods, 1990) and possibly also the mechanical properties of the cured adhesive.

PUR adhesives are still matter of basic research focusing on their properties along the wood bond line and penetration into the wood (Gindl et al., 2005; Müller et al., 2005). Widsten et al. (2006) investigated the factors influencing timber gluability with 1C PUR on nine Australian timber species. The penetration of adhesives into wood, including PUR resins, was in focus of the paper of Konnerth et al. (2008). The wetting of modified wood with adhesives, including PUR, was tested by Bryne and Wålinder (2010).

The properties of 1C PUR adhesives are also in focus of the present paper because of the large importance of this type of adhesives. Their production takes place in a two-stage process: (1) In the first stage, an exothermic polyaddition reaction of polyol with an excess of polyisocyanate leads to urethane prepolymers with a defined amount of free monomeric polyisocyanate, which acts as a solvent for the prepolymer. (2) In the second stage, the prepolymer is mixed with different additives to generate the final product. The properties of the final product can largely be controlled by varying

Table 3.3.1 Chemical and physical parameters of laboratory polyurethane prepolymers and adhesives

	Prepolymers			Formulated adhesives		
PUR code	P1	P2	P3	A1	A2	A3
NCO content [% w/w]	16	16	16	16	16	16
Cross-link density [mol kg^{-1}]	0.22	0.87	1.25	0.22	0.87	1.25
Urea group content [mol kg^{-1}]	2	2	2	2	2	2
Urethane group content [mol kg^{-1}]	0.58	0.77	0.58	0.58	0.77	0.58
Viscosity (23°C) [mPa s]	1310	8920	5460	2500	19000	14000
Film drying time [min]	960	660	720	180	160	180
Defoamer[1]	-	-	-	yes	yes	yes
Pyrogenic silica[1]	-	-	-	yes	yes	yes
Amine catalyst[1]	-	-	-	yes	yes	yes

[1]Identical amounts of defoamer, pyrogenic silica and catalyst for all adhesives

not only by the molar ratio of the components but also by additives. Several details are not well known in this regard.

Thus the objective of this study is to investigate the influence of the adhesive formulation on the mechanical properties of the cured adhesive. Of particular interest are effects caused by the addition of amin catalyst and additives such as pyrogenic silica and defoamers. This study is aimed at clarifying whether a catalyst accelerates side-reactions that might have negative effects on the mechanical properties. The question should also be answered whether the adjustment of viscosity leads to agglomerations of the pyrogenic silica within the prepolymer matrix. Furthermore, the influence of softening effects by surfactants (defoamers) should be investigated.

3.3.3 Material and methods

Three prepolymers were produced by Bayer MaterialScience (Leverkusen, Germany) based on the reaction of isocyanates and polyols. The isocyanates used were mixtures of methylene diphenyl diisocyanate (MDI) monomer and polymeric MDI. Since polymeric MDI contains more than two isocyanate groups per molecule, the resulting overall functionality of the prepolymers is >2. The polyols used resulted from a base-catalyzed reaction of 1,2-propylene glycol with propylene oxide. The water-free polyol component was mixed with the isocyanate components at 50 to 70°C under continuous stirring, until the desired NCO content was reached and remained constant. Three different functionality adjustments were chosen, each having an NCO content of about 16 % (Table 3.3.1).

These prepolymers formed the basis for the formulation of adhesives by adding and dispersing identical amounts of defoamer, pyrogenic silica and amine catalyst by the adhesive producer Purbond (Sempach-Station, Switzerland). A similar reactivity of the adhesives was achieved, measured as open time in the range of 60 to 90 min. Additionally, commercially available adhesives for structural wood bonding

Table 3.3.2 Bonding parameters of the tested adhesives and prepolymers

Adhesive	Adhesive / hardener ratio[a]	EMC[b] [%]	Pressure [MPa]	Pressing time	Application	Spread [g m^{-3}]
MUF	100/35	~12	0.8 - 1.2	4	2 sides	200
PRF	100/20	~12	0.8 - 1.2	5	2 sides	180
1C PUR commercial	-	≥8	0.6 - 1.0	3	1 side	200
1C PUR prepolymer	-	-	-	12	1 side	200
1C PUR adhesive	-	-	-	6	1 side	200

[a] applied in liquid state
[b] equilibrium moisture content as recommended by adhesive producer

– phenol-resorcinol-formaldehyde (PRF), melamine-urea-formaldehyde (MUF), and 1C PUR – were included into the study for comparison (for tensile shear results of these and further adhesives see Clauß, Joščák and Niemz (2011)). The adhesive performances were tested by: (1) Longitudinal tensile shear strength according to DIN EN 302-1 (2004). (2) Tensile properties of adhesive films according to DIN EN ISO 527-1 (1996). (3) Micro-mechanical properties by means of nanoindentation.

In accordance with DIN EN 302-1 (2004), beech (*Fagus sylvatica* L.) was selected as substrate because of (1) its low content of extractives and (2) its higher strength compared with spruce, which is common in timber engineering. The raw density of the beech was 735 ± 34 kg m^{-3} at an equilibrium moisture content of 12 ± 1 %. Pressure of about 0.8 MPa was applied for specimen preparation and subsequent curing under standard climatic conditions (20°C, 65 % RH) was performed at least for 7 days. The specific bonding parameters for the different adhesives are listed in Table 3.3.2.

The bonds were tested under tensile shear load (standard climatic conditions, Zwick/Roell Z010 universal testing machine), which implies a position-controlled measurement of the load-displacement curves of adhesively bonded lap joints until failure. The strain measurement was performed by means of a clip-on displacement transducer. The wood failure percentage (WFP) was estimated visually in steps of 10 %.

Cubes with an edge length of about 10 mm were cut from the climatically conditioned joints for nanoindentation. The bond line of the specimens was oriented in the middle. Subsequently, a gently sloping apex was microtomed on the surface of the unembedded specimens. Afterwards, a tip was sliced off the apex with a diamond knife microtome similar to the preparation method described in detail by Jakes et al. (2008).

A Hysitron Triboindenter (Hysitron, Minneapolis, MN) equipped with a three-faceted diamond pyramid (Berkovich) indenter tip was used for the measurements. The machine compliance was determined by a series of indents in the center of a fused silica standard with loads ranging from 0.05 to 10.00 mN. From the load-depth graph, hardness and Young's modulus were calculated according to Oliver and Pharr (1992). Data of Young's modulus were not corrected (Gindl et al., 2004). The deformation energy was determined as the integral of load vs. displacement as previously described (Konnerth, Gindl and Müller, 2007; Stanzl-Tschegg, Beikircher and Loidl, 2009).

The indents (10 indents spaced 5 µm apart) were placed along the bond line according to Fig. 3.3.3(a). Under these circumstances an interaction between the indents can be disregarded (Jakes et al., 2008). The experiments were performed in load-controlled mode (test speed $1.7\,\mathrm{N\,s^{-1}}$, peak force 400 µN) resulting in indent depths of about 450 nm for the cured adhesive and 200 nm for the secondary cell wall, respectively. At peak load, a hold period of 5 s was included in order to determine the viscoelastic performance. For the preparation of the adhesive films for the tensile tests, the liquid adhesives were applied to a plastic sheet. The typical foaming effect of polyurethanes could be minimized by applying a film thickness of only about 0.25 mm in a 50 % RH environment. The films were peeled from the plastic sheet and stored for a minimum of 3 days under standard climatic conditions. Specimens of shape type 1B according to DIN EN ISO 527-3 (2003) were prepared from these films.

The tensile properties of the adhesives were obtained according to DIN EN ISO 527-1 (1996) (standard climatic conditions, Zwick/Roell Z100 universal testing machine, 500 N load cell, test speed $5\,\mathrm{mm\,min^{-1}}$). The strain was measured optically by a video-extensometer. Thereby tensile and transverse deformation was recorded; Young's modulus, strength and strain at maximal load, and the Poisson's ratio were calculated from the load displacement curves. The mean values presented in Table 3.3.3 are a series of at least six specimens.

FTIR spectra of the cured adhesives were recorded on a Thermoelectron Continuum FTIR Microscope in transmission mode using thin sections taken from the interface region of the bond line sliced with a microtome knife. Spectra were taken from an area of a size of an adhesively filled vessel. The bonds had been stored for one month under standard climatic conditions and the spectra were recorded directly after the thin sections had been prepared. The measurements were repeated on the same specimens after 1 and 4 weeks. As a reference, spectra were also recorded from the adhesives in a liquid state by attenuated total reflectance (ATR) spectroscopy. The ratio between the NCO band at $2278\,\mathrm{cm^{-1}}$ and the stable aromatic C-H band at $3030\,\mathrm{cm^{-1}}$, which remains constant during the reaction, was calculated and taken as a measure for free NCO groups.

For a qualitative analysis of the bond line and interphase region, micrographs were taken by means of a Dual Beam Scanning Electron Microscope (Quanta 200 3D, FEI) in low vacuum mode. Additional AFM imaging was performed in tapping mode by means of the Atomic Force Microscope (Digital Instruments D3000, Bruker) providing phase contrast images of the polymer structure.

3.3.4 Results

Tensile shear strength on bonded wood joints

The tensile shear test on bonded wood joints revealed significant differences between the tested prepolymers and the formulated adhesives (Fig. 3.3.1(a)). Each type of adhesive reached, on average,

3 Main investigations

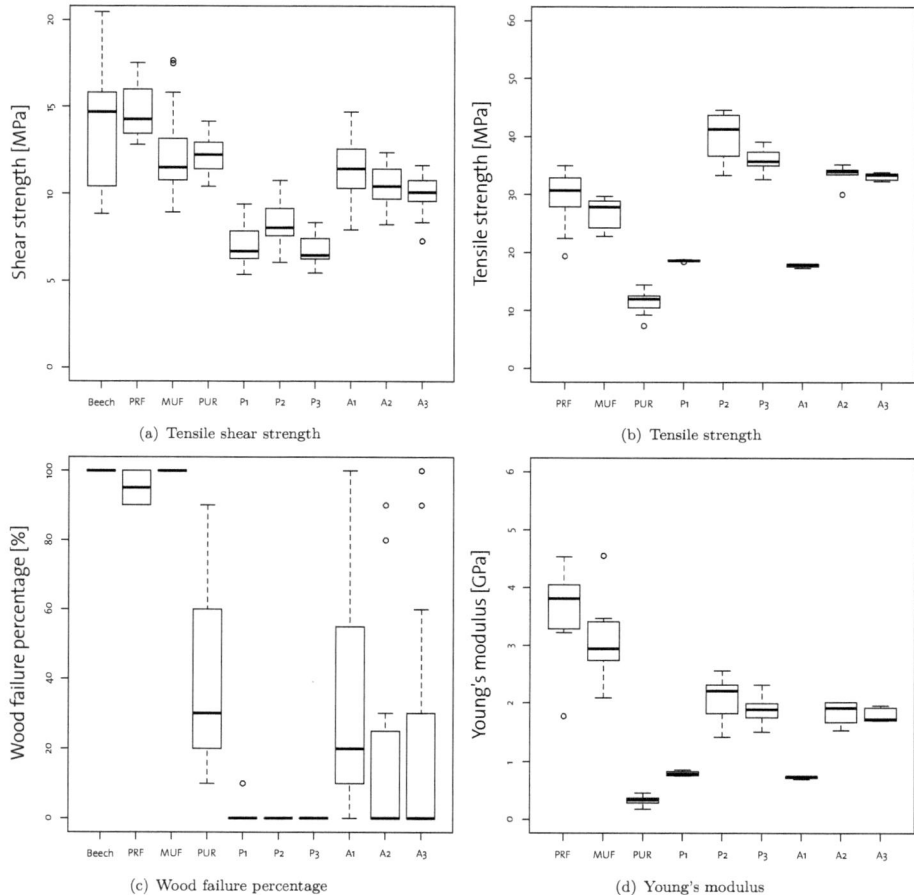

Fig. 3.3.1 Mechanical properties of bonded wood joints and films

40 % higher values than the basic prepolymer. Due to the higher strength of the formulated adhesives, wood failure partially occurred. The mean WFP for all 1C PUR adhesives was below 50 % (Fig. 3.3.1(c)). The prepolymers, however, showed no wood failure at all, irrespective of their cross-link density.

The tensile shear strength with the commercial 1C PUR adhesive was on average about 15 % higher than with the laboratory formulations; the WFP was about 40 %. PRF reached significantly higher values than all the other adhesives and its strength was comparable with that of beech wood. MUF and the commercial 1C PUR exhibited similar values within the interquartile range of wood strength; however, the WFP of PUR was considerably lower. Both polycondensation resins showed wood failure but not cohesive adhesive failure.

Tensile properties of adhesive films

The tensile strengths (Fig. 3.3.1(b)) and the Young's moduli (Fig. 3.3.1(d)) were significantly different between the adhesive types and within the particular groups of prepolymers and formulated adhesives. The prepolymers with medium P2 and high cross-link density P3 reached tensile strengths and Young's moduli that were about twice as high as those of the prepolymer with low cross-link density P1.

The adhesives formulated from these prepolymers resulted in nearly the same tensile strengths and Young's moduli of the formed films as the respective prepolymers. The mean values of the groups (prepolymer and formulated adhesive) did not differ significantly, as shown by the Welch two-sample t-test ($\alpha = 0.05$). The difference between the commercial products and the laboratory formulations is also significant. The commercial 1C PUR adhesive had the lowest tensile strength (11.5 MPa) and also the lowest Young's modulus (329 MPa) of all tested films.

The strengths of both polycondensation resins were similar to the results of the laboratory 1C PUR adhesives, with PRF showing a slightly higher strength (29.5 MPa) compared with MUF (26.7 MPa). Due to the brittle behavior of the films, the fracture occurred spontaneously without developing a yield point; also the Young's modulus was considerably higher compared with the 1C PUR adhesives. PRF reached the highest Young's modulus with 3.6 GPa, followed by MUF with 3.1 GPa.

Nanoindentation

The prepolymers differed significantly regarding the parameters determined by nanoindentation. The prepolymer with the lowest cross-link density P1 exhibited significantly lower hardness (Fig. 3.3.2(a)) and Young's modulus (Fig. 3.3.2(c)) than the prepolymers with medium P2 and high P3 cross-link density. A correlation between cross-link density and mechanical properties, however, could not be detected. In spite of the higher cross-link density, P3 resulted in a lower hardness and stiffness than P2. The deformation energy (Fig. 3.3.2(e)) showed the opposite tendency. As expected, the prepolymer with the lowest cross-linking was able to absorb the most energy.

By means of nanoindentation it was possible to distinguish indents in phases of the bond line with preferably hard and soft segment-rich phases of the bond line. The hard segment domain of P1 showed significantly higher values for indent hardness and Young's modulus. The deformation energy in contrast was lower compared to the soft segment domain. The prepolymers with higher cross-link density did not show segregations in the polymer structure. The formulated adhesives showed a similar picture; however, differences in hardness, Young's modulus, and deformation energy between

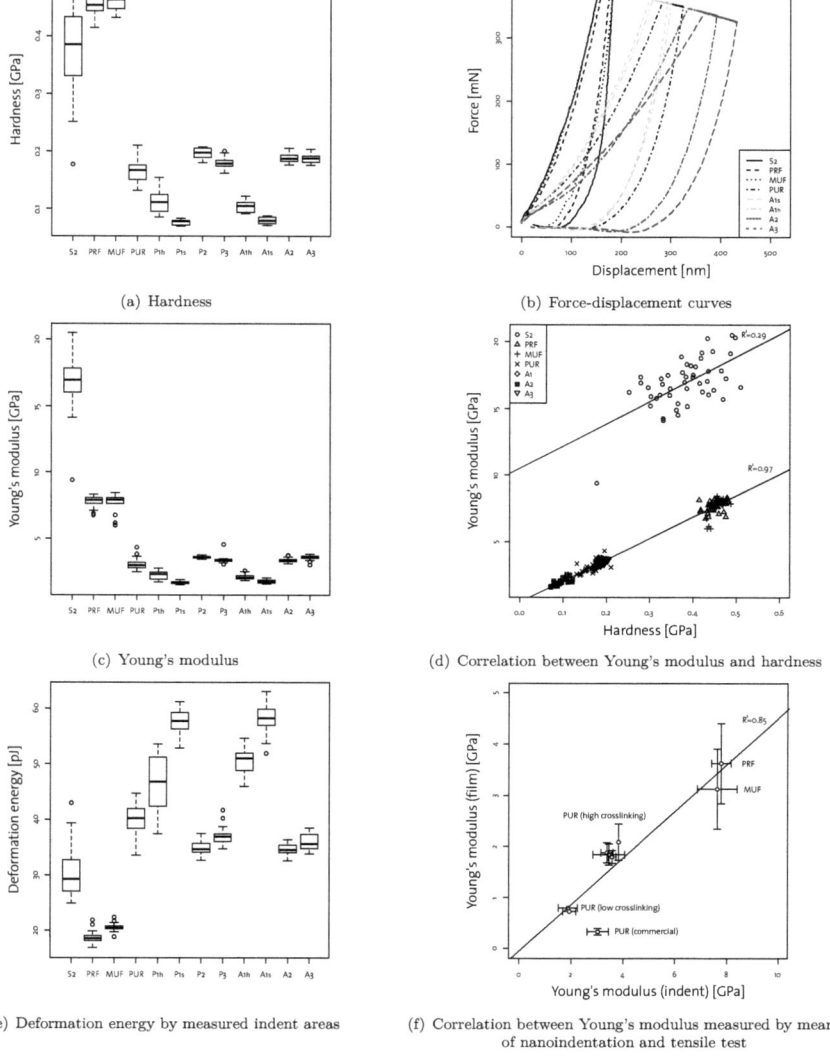

Fig. 3.3.2 Nanoindentation results of the bond line interphase region. Commercial adhesive: PRF, MUF, 1C PUR. PUR prepolymer: P1, P2, P3. Laboratory formulated 1C PUR adhesive: A1, A2, A2. (h) hard and (s) soft segment-rich phase. (S2) cell wall of beech wood. Error bar: std. deviation. Box: quartile of distribution. Whisker: 1.5 × interquartile range. Point: outliers.

the adhesives A3 (high cross-link density) and A2 (medium cross-link density) were not statistically significant, as this was also proven by means of an analysis of variance ($\alpha = 0.05$).

The values for the commercial 1C PUR were in between the laboratory formulations with low and medium cross-link density. PRF and MUF showed no significant difference in hardness and stiffness; MUF was able to absorb more deformation energy than PRF. For both adhesives hardness and stiffness were clearly higher than for all 1C PUR adhesives. The hardness of the cell wall of beech was slightly lower than for the amino- and phenoplastic resins; however, the variance for the wood samples was much higher than that of the two adhesives. The Young's modulus of beech was considerably higher than the values for the adhesives. The force-displacement curves in Fig. 3.3.2(b) illustrate the amount of deformation energy absorbed, which is represented by the area below the curve. The curves for the polycondensation resins were similar to those of the wood cell wall, whereas the PUR formulations are characterized by a minor slope and a significantly greater area below the curve. The creep deformation, which can be estimated by the displacement under constant maximum load was also higher for PUR adhesives compared to the polycondensation resins and the cell wall.

3.3.5 Discussion

Prepolymer composition

The variation in the prepolymer configuration is reflected by the tensile tests on films and nanoindentation results in the interphase region. A low cross-link density of the prepolymer resulted in lower stiffness, strength, and hardness. Caused by the minor cross-linking, the alternating hard and soft segments of the molecule chains could easily segregate to form hard segment domains in the soft segment matrix. By means of AFM phase contrast images (Fig. 3.3.3(b)), it was possible to visualize these domains due to their differing micromechanical properties. Phase contrast imaging shows a brighter contrast of areas in the sample with higher elasticity and/or lower adhesive forces between the silicon tip of the AFM cantilever and the sample material. At higher cross-link density, the structure of the prepolymer was significantly more homogeneous.

The tensile shear strength of bonded wood joints is in contrast to the results of pure prepolymers not affected by their chemical composition, although it might be claimed that in the case of no wood failure the shear strength of the joint completely relies on the cohesive strength of the adhesive. The mechanical properties are negligible in case of a poor bond line formation of the prepolymers. This is clearly identifiable by comparison of ESEM micrographs showing bonds of beech wood with either prepolymer (Fig. 3.3.3(c)) or formulated adhesive (Fig. 3.3.3(d)). Vessels in a distance of more than 500 µm from the bond line were filled by the prepolymer and the bond line showed several imperfections. A quantification of the lower pore space saturation by means of synchrotron tomography confirmed a starved bond line (lacking of resin) in the case of PUR prepolymers (Haß et al., 2012). A satisfactory bond formation is hardly possible under these circumstances.

3 Main investigations

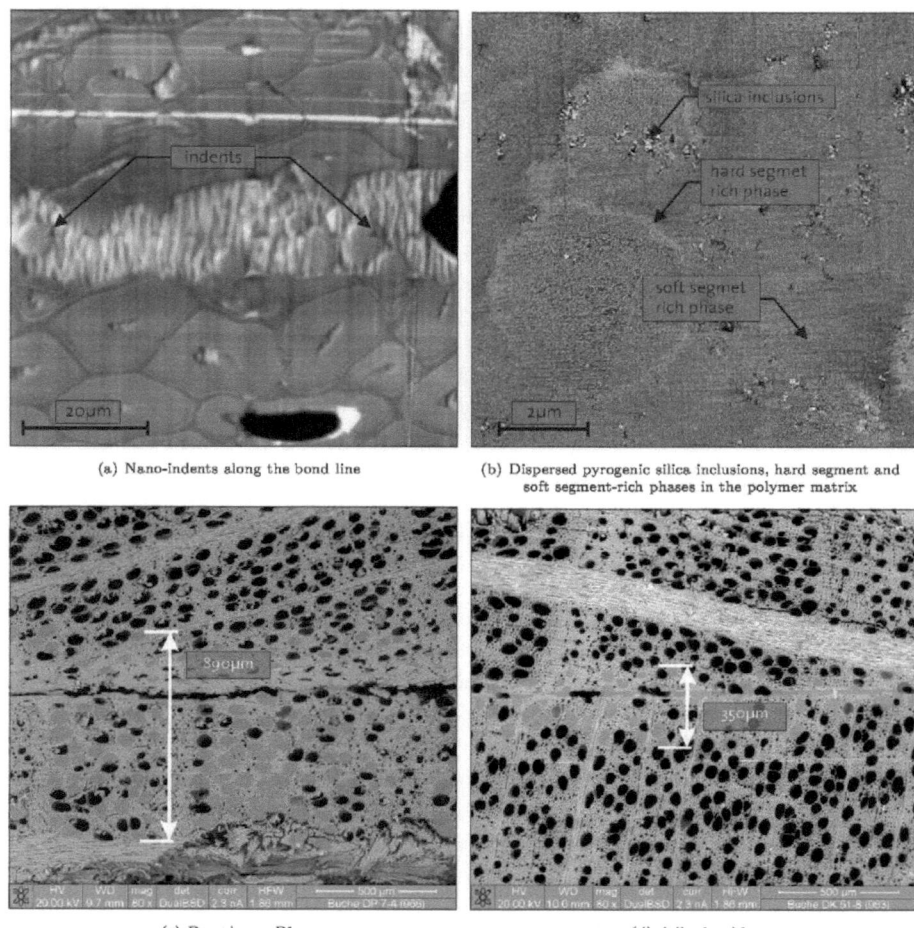

(a) Nano-indents along the bond line

(b) Dispersed pyrogenic silica inclusions, hard segment and soft segment-rich phases in the polymer matrix

(c) Prepolymer P1

(d) Adhesive A1

Fig. 3.3.3 AFM and ESEM micrographs of bond lines show interphases formed by penetration of prepolymer and formulated adhesive A1 into wood

Adhesive formulation

Rheology, kinetics, and wettability are mainly influenced by the formulation of the adhesive. The penetration of the adhesives is considerably limited due to the adjustment of these properties. Since the prepolymers P2 and P3 had considerably higher viscosities than adhesive A1, a correlation be-

Table 3.3.3 Summary of the results determined by tensile test on films, nanoindentation and tensile shear test of bonded wood joints

Adhesive		E_{film} [GPa]	σ_{film} [MPa]	ε [%]	ρ [-]	E_{indent} [GPa]	H [GPa]	W [pJ]	t [MPa]	WFP [%]
Wood/S2		-	-	-	-	16.72	0.38	30.2	13.97	100
	v [%]	-	-	-	-	12	18	14	29	-
PRF		3.63	29.5	0.86	-	7.78	0.45	18.5	14.72	100
	v [%]	22	17	14	-	5	4	5	14	-
MUF		3.13	26.65	1.09	-	7.63	0.46	20.5	12.25	100
	v [%]	25	11	22	-	10	3	4	20	-
PUR		0.33	11.5	18.83	0.45	3.03	0.16	40	12.15	40
	v [%]	20	15	13	18	14	12	6	8	-
P1		0.8	18.64	6.28	0.48	1.88	0.09	53.2	6.93	0
	v [%]	5	1	4	10	20	27	13	16	-
P2		2.09	40.03	3.47	0.4	3.83	0.2	34.7	8.28	0
	v [%]	17	10	7	18	3	4	3	15	-
P3		1.88	35.17	3.54	0.42	3.38	0.18	36.7	6.73	0
	v [%]	10	5	13	16	7	5	4	12	-
A1		0.73	17.84	6.57	0.45	1.94	0.09	53.8	11.29	40
	v [%]	3	2	3	5	13	16	8	17	-
A2		1.84	33.55	2.76	0.43	3.45	0.19	34.5	10.46	20
	v [%]	11	5	8	14	18	9	3	13	-
A3		1.79	33.19	3.16	0.38	3.58	0.19	35.8	10.07	20
	v [%]	7	2	12	21	4	4	4	12	-

E_{film}, Young's modulus measured by tensile test; σ_{film}, tensile strength; ε, strain at failure; μ, Poisson ratio; E_{indent}, Young's modulus measured by nanoindentation; H, hardness; W, deformation energy; τ, tensile shear strength; WFP, wood failure percentage; $\bar{x}$, mean value; ν, coefficient of variation

tween viscosity and shear strength cannot be established by the current results. Previous studies by Clauß et al. (2011b) and Haß et al. (2012) do not provide any evidence of this either. The notably different film drying times (Table 3.3.1) of prepolymers and adhesives quantify their different reaction rates. The faster reaction of the adhesive's NCO groups with water, accelerated by the addition of catalyst, also restricts the penetration and therefore promotes greater concentration of the adhesive in the bond line. However, it might also be possible that NCO groups of prepolymers do not fully react with water and therefore the network of the prepolymer is less cross-linked compared with a catalyzed adhesive.

FTIR spectroscopy of the cured adhesives showed significant differences in the absorption peaks due to the NCO groups that appear as intense and sharp bands at about 2278 cm^{-1} in the ATR spectra (Fig. 3.3.4(a)). After curing under standard climatic conditions, the NCO group absorption of all of the bonds was reduced drastically but did not disappear completely (Fig. 3.3.4(b)). In the case of the formulated adhesive, the absorbance is comparatively low amounting to a free NCO content of 0.04 %. A repeated measure of the same thin section after 6 days revealed that the peak disappeared. In the

3 Main investigations

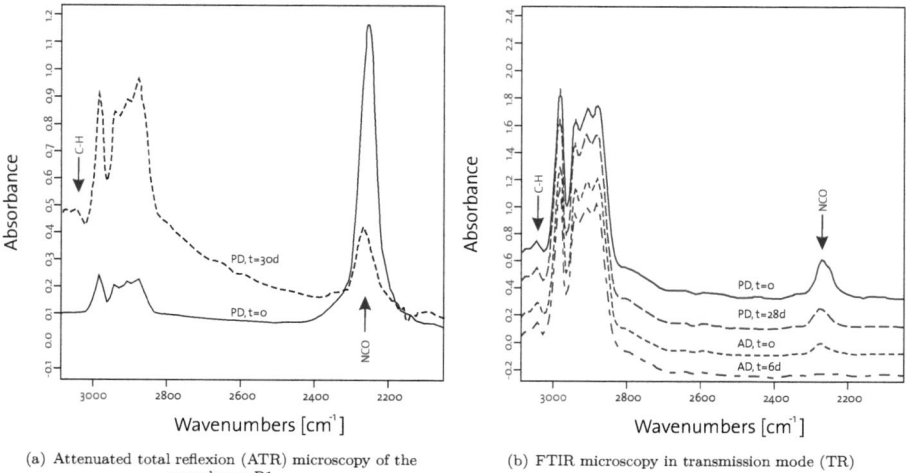

(a) Attenuated total reflexion (ATR) microscopy of the prepolymer P1

(b) FTIR microscopy in transmission mode (TR)

Fig. 3.3.4 Infrared spectra of liquid adhesive and thin sections cut from the interphase region of the bond line of samples bonded by prepolymer P1 and adhesive A1

prepolymer, an NCO content of about 0.13 % was measured; after one month only 0.06 % of unreacted NCO was detected, compared with 16 % NCO in the liquid state. It may therefore be assumed that the cross-link reaction was almost complete.

The formulation of the prepolymer to an adhesive led to the formation of a well-defined bond line thickness, resulting in significantly higher tensile shear strength of the formulated adhesives compared to the prepolymers. As the mechanical properties measured by tension test and nanoindentation are nearly identical for prepolymers and formulated adhesives, one can conclude that the cohesion of the adhesive is almost completely governed by the prepolymer configuration. Negative side effects due to the addition of catalyst, defoamer, and pyrogenic silica could not be detected. Nonetheless, it cannot be ruled out that side reactions occur; NMR spectroscopy might provide objective evidence of this (Ni and Frazier, 1998).

The Young's modulus obtained with the tension test on thin films was (for all tested adhesives) considerably lower compared to the values measured by nanoindentation (Fig. 3.3.2(f)); similar findings were also reported by Lucas, Rosenmayer and Oliver (1998) and Konnerth et al. (2006). Several basic differences in the methods used have a strong influence on the test results. The three dimensional stress state below the indenter tip is affected by shear forces in comparison with the uniaxial tension test. The sample area can be considered to be different by several orders of magnitude, thus local imperfections (e.g., CO_2 blisters or pyrogenic silica) have a different impact on the results. Additionally, the contact depth, geometry of the indenter tip and unloading speed are important influencing factors of the nanoindentation method (VanLandingham et al., 2001). The results of the tension tests are influenced by the strain rate and the type (secant, regression) and range ($\sigma_1 - \sigma_2$, (DIN EN ISO 527-1, 1996)) of Young's modulus determination. The high linear correlation ($R^2 = 0.84$) between the two methods, however, confirms that the differentiation of the polymers is independent of the method used, since different methods and experimental parameters affect the absolute values, but scarcely the relation of the adhesives to each other.

Commercial adhesives

The commercial 1C PUR revealed good bond properties despite the low stiffness and strength of the pure adhesive, which ranged below the values of the laboratory formulations. Obviously, suitable rheological and film formation properties are more important for good bond quality than the mechanical properties of the tested adhesives. The deviation from the high correlation between Young's moduli obtained by tensile test and nanoindentation (Fig. 3.3.2(f)) suggests an underestimated Young's modulus calculated from the tensile test; however, other studies report very similar results for this adhesive (Konnerth et al., 2006) A possible reason could have been the fibrous filler material of the commercial PUR, which tended to agglomerate during the film formation and was also aligned in the direction of application. The filler material, however, is largely responsible for the better performance in the shear test (Clauß et al., 2011a).

In contrast to PUR, amino- and phenoplastic resins are characterized by high stiffness and brittle failure behavior; however, both adhesive types are able to reach high tensile shear strengths. The high WFP for PRF and MUF demonstrates that the wood strength is clearly exceeded by these adhesives, thus properties of the adherend determined the resulting values (Bergman et al., 2010). It is known that PUR adhesives reach lower WFP at equal shear strength (Niemz and Allenspach, 2009; Clauß, Joščák and Niemz, 2011), in particular under wet conditions (Vick and Okkonen, 1998). Possible reasons for this could be changes in wood properties due to inter- or intracellular adhesive penetration (Konnerth and Gindl, 2006; Konnerth, Valla and Gindl, 2007; Stöckel et al., 2010) or gross penetration through the cellular network (Kamke and Lee, 2007). The high gross penetration of the ductile polyurethane possibly contributes to delaying wood failure by reinforcing the wood. Furthermore, intracellular penetration of PRF or MUF in the interphase region could promote wood failure due to embrittlement of the cell wall. However, more experiments are needed to test this hypothesis.

Impact on the wood

The current nanoindentation measurements on wood cell walls do not allow conclusions about changes in their mechanical properties due to a possible penetration of the adhesive into the intercellular regions. If penetration still occurred, beech (in contrast to spruce tested in other publications) is most likely less affected as the Young's modulus obtained by indentation is significantly higher compared to the adhesives tested. The high variation in the values determined for wood can be explained (1) by specimens taken from varying samples and (2) indents in different cell types (fiber tracheids and vessels) with varying micro-fibril angles, which are highly correlated to the stiffness of the material (Gindl and Schöberl, 2004; Tze et al., 2007; Donaldson, 2008; Wu et al., 2009).

Young's modulus and hardness of the adhesives showed a high linear correlation ($R^2 = 0.97$) (Fig. 3.3.2(d)) also documented by Shi et al. (2006). The cell wall of beech shows a similar trend but worse correlation ($R^2 = 0.29$); however, the moduli are higher by about 10 GPa compared to the various adhesive films. This shows again that beech exhibits in the area of S2 cell wall layer a hardness in the range of PRF or MUF but with a considerably stiffer behavior.

3.3.6 Conclusions

- The mechanical properties of the 1C PUR adhesives are determined by the prepolymer compositions.

- The adjustment of viscosity and reactivity significantly enhanced the bonding performance of formulated adhesives in comparison with unformulated prepolymers.

- Negative side effects due to the addition of catalyst, pyrogenic silica, defoamer, etc., on the mechanical properties or the bonding performance could not be detected.

- Too low cross-linking of the polyurethane prepolymer can result in segregation into hard and soft segment-rich phases of the polymer with differing mechanical properties.

- 1C PUR adhesives are characterized by significantly lower stiffness and hardness compared to amino- and phenoplastic resins, but absorb much more deformation energy and show ductile failure behavior leading to lower wood failure.

References

Bergman, R. et al. (2010) Wood handbook: Wood as an engineering material. Madison, WI: Forest Products Laboratory, 508 p.

Bryne, L.E. and **Wålinder, M.E.P. (2010)** Ageing of modified wood. Part 1: Wetting properties of acetylated, furfurylated, and thermally modified wood. Holzforschung, 64 No. 3, 295–304

Clauß, S. et al. (2011a) Improving the thermal stability of one-component polyurethane adhesives by adding filler material. Wood Sci. Technol. 45 No. 2, 383–388

Clauß, S. et al. (2011b) Influence of the chemical structure of PUR prepolymers on thermal stability. Int. J. Adhes. Adhes. 31 No. 6, 513–523

Clauß, S., Joščák, M. and **Niemz, P. (2011)** Thermal stability of glued wood joints measured by shear tests. Eur. J. Wood Prod. 69 No. 1, 101–111

DIN EN 302-1 (2004) Adhesives for load-bearing timber structures – Test methods – Part 1: Determination of bond strength in longitudinal tensile shear strength. Berlin: Beuth Verlag

DIN EN ISO 527-1 (1996) Plastics – Determination of tensile properties – Part 1: General principles. Berlin: Beuth Verlag

DIN EN ISO 527-3 (2003) Plastics – Determination of tensile properties – Part 3: Test conditions for films and sheets. Berlin: Beuth Verlag

Donaldson, L. (2008) Microfibril angle: Measurement, variation and relationships - A review. IAWA Journal, 29 No. 4, 345–386

Gindl, W. et al. (2004) Mechanical properties of spruce wood cell walls by nanoindentation. Appl. Phys. A: Mater. Sci. Process. 79 No. 8, 2069–2073

Gindl, W. and **Schöberl, T. (2004)** The significance of the elastic modulus of wood cell walls obtained from nanoindentation measurements. Compos. Part A, 35 No. 11, 1345–1349

Gindl, W. et al. (2005) Direct measurement of strain distribution along a wood bond line. Part 2: Effects of adhesive penetration on strain distribution. Holzforschung, 59 No. 3, 307–310

Haß, P. et al. (2012) Adhesive penetration in beech wood: Experiments. Wood Sci. Technol. 46 No. 1-3, 243–256

Hse, C.Y. and **Kuo, M.I. (1988)** Influence of Extractives On Wood Gluing and Finishing - A Review. Forest Prod. J. 38 No. 1, 52–56

Ionescu, M. (2005) Chemistry and technology of polyols for polyurethanes. Shrewsbury: Smithers Rapra Press, 602 p.

Jakes, J.E. et al. (2008) Experimental method to account for structural compliance in nanoindentation measurements. J. Mater. Res. 23 No. 4, 1113–1127

Kamke, F.A. and **Lee, J.N. (2007)** Adhesive penetration in wood - A review. Wood Fiber Sci. 39 No. 2, 205–220

Konnerth, J. and **Gindl, W. (2006)** Mechanical characterisation of wood-adhesive interphase cell walls by nanoindentation. Holzforschung, 60 No. 4, 429–433

Konnerth, J., Gindl, W. and **Müller, U. (2007)** Elastic properties of adhesive polymers. I. Polymer films by means of electronic speckle pattern interferometry. J. Appl. Polym. Sci. 103 No. 6, 3936–3939

Konnerth, J. et al. (2008) Adhesive penetration of wood cell walls investigated by scanning thermal microscopy (SThM). Holzforschung, 62 No. 1, 91–98

Konnerth, J. et al. (2006) Elastic properties of adhesive polymers. II. Polymer films and bond lines by means of nanoindentation. J. Appl. Polym. Sci. 102 No. 2, 1234–1239

Konnerth, J., **Valla, A.** and **Gindl, W.** (**2007**) Nanoindentation mapping of a wood-adhesive bond. Appl. Phys. A: Mater. Sci. Process. 88 No. 2, 371–375

Lucas, B.N., **Rosenmayer, C.T.** and **Oliver, W.C.** (**1998**) Mechanical characterization of submicron polytetrafluoroethylene (PTFE) thin films. In Materials Research Society symposia proceedings. Warrendale, PA: Materials Research Society, 97–102

Müller, U. et al. (**2005**) Direct measurement of strain distribution along a wood bond line. Part 1: Shear strain concentration in a lap joint specimen by means of electronic speckle pattern interferometry. Holzforschung, 59 No. 3, 300–306

Ni, J.W. and **Frazier, C.E.** (**1998**) N-15 CP/MAS NMR study of the isocyanate/wood adhesive bondline. Effects of structural isomerism. J. Adhes. 66 No. 1-4, 89–116

Niemz, P. (**1993**) Physik des Holzes und der Holzwerkstoffe. Leinfelden-Echterdingen: DRW-Verlag, 243 p.

Niemz, P. and **Allenspach, K.** (**2009**) Studies on the influence of temperature and timber moisture on the failure behaviour of selected adhesives under tensile shear load. Bauphysik, 31 No. 5, 296–304

Oliver, W.C. and **Pharr, G.M.** (**1992**) An improved technique for determining hardness and elastic modulus using load and displacement sensing indentation experiments. J. Mater. Res. 7 No. 6, 1564–1583

Shi, S.Q. et al. (**2006**) Characterization of engineering wood adhesive behavior treated at elevated temperatures. In **Friehard, C.R., editor:** Wood Adhesives 2005. Madison, WI: Forest Products Society, 163–169

Siau, J.F. (**1984**) Transport processes in wood. Berlin: Springer, Springer series in wood science, 245p.

Stanzl-Tschegg, S., **Beikircher, W.** and **Loidl, D.** (**2009**) Comparison of mechanical properties of thermally modified wood at growth ring and cell wall level by means of instrumented indentation tests. Holzforschung, 63 No. 4, 443–448

Stöckel, F. et al. (**2010**) Tensile shear strength of UF- and MUF-bonded veneer related to data of adhesives and cell walls measured by nanoindentation. Holzforschung, 64 No. 3, 337–342

Suomi-Lindberg, L., **Pulkkinen, P.** and **Nussbaum, R.** (**2002**) Influence of the wood component on the bonding process and the properties of wood products. In **M. Dunky, A. Pizzi, M. van Leemput, editor:** COST Action E13 - State-of-the-art report - Wood adhesion and glued products., 121–143

Tze, W.T.Y. et al. (**2007**) Nanoindentation of wood cell walls: Continuous stiffness and hardness measurements. Compos. Part A, 38 No. 3, 945–953

VanLandingham, M.R. et al. (**2001**) Nanoindentation of polymers: An overview. Macromol. Symp. 167 No. 1, 15–43

Vick, C.B. and **Okkonen, E.A.** (1998) Strength and durability of one-part polyurethane adhesive bonds to wood. Forest Prod. J. 48 No. 11-12, 71–76

Widsten, P. et al. (2006) Factors influencing timber gluability with one-part polyurethanes, studied with nine Australian timber species. Holzforschung, 60 No. 4, 423–428

Woods, G. (1990) The ICI polyurethanes book. New York: John Wiley, 362p.

Wu, Y. et al. (2009) Use of nanoindentation and silviscan to determine the mechanical properties of 10 hardwood species. Wood Fiber Sci. 41 No. 1, 64–73

3.4 Paper IV

Wood Sci. Technol. (2011) 45:383-388

Improving the thermal stability of one-component polyurethane adhesives by adding filler material

Sebastian Clauß[1], Karin Allenspach[1], Joseph Gabriel[2] and Peter Niemz[1]

[1]Institute for Building Materials, ETH Zurich, Schafmattstrasse 6, 8093 Zurich, Switzerland
[2]Purbond AG, Industriestrasse 17a, 6203 Sempach-Station, Switzerland

3.4.1 Abstract

The aim of the current study is to improve the thermal stability of one-component moisture-curing polyurethane adhesives. Our approach tends to add suitable filler materials to the adhesive and to study the resulting effects. The investigation covers mechanical tests to determine the shear strength of the bonded wood joints according to DIN EN 302-1 (2004). Furthermore, the distribution of the filler material within the adhesive is shown by means of environmental scanning electron microscopy combined with energy dispersive X-ray spectroscopy analysis. The thermal stability of the bonded wood joints could be significantly improved by adding chalk with a volume fraction of 30 % to the adhesive.

3.4.2 Introduction

One-component polyurethane adhesives (1C PUR) are increasingly used for the bonding of wood. The properties of the reacted polymers (as elasticity, strength, temperature and moisture resistance) are influenced by the prepolymer as well as by additives like surfactant, catalyst and especially filler material. Filler materials are non-volatile, non-gluing matters which are insoluble in the adhesive. Common fillers are fibres (glass fibre, mica), powders (cellulose, aluminium oxide, silica), sheet-like materials (talc), cubic materials (chalk, barytes) (Zeppenfeld and Grunwald, 2005) or nowadays nanoparticles (Park, Kim and Lee, 2009) or functionalized nanoclays (Dodiuk et al., 2006).

In the past, several investigations on different types of adhesives and fillers have been carried out. The mechanical properties of polyvinyl acetate, depending on morphology and chemical structure of the filler material (calcium carbonate), were investigated by Kovačević et al. (1996). The influence of the same filler on the rheological and adhesion properties of a water-based polyurethane dispersion was investigated by Muñoz-Milán et al. (2005). Mansouri and Pizzi (2007) improved the

3 Main investigations

Table 3.4.1 Adhesives' structural properties

Adhesive	A	B	C
Filler content [%]	0	15	30
Isocyanat [%]	15	15	15
Open time [min]	60	60	60
Viscosity [mPas]	6580	9340	13960

performance of urea-formaldehyde and phenol-formaldehyde resin by adding micronized polyurethane powder. Sepulcre-Guilabert, Ferrándiz-Gómez and Martín-Martínez (2001) proposed natural ultramicronized calcium carbonate and mixtures of fumed silica with natural ultramicronized calcium carbonate as filler for solvent-based polyurethane adhesives.

Investigations on the structure-property relationships of 1C PUR adhesives for wood, including adhesives with fibrous fillers, and their sensitivity to low wood moisture content (WMC) were carried out by Beaud, Niemz and Pizzi (2006). In contrast, Richter and Schierle (2003) and Schrödter and Niemz (2006) investigated the adhesive performance of 1C PUR under high moisture and temperature conditions. It can be concluded that the bonding strength of 1C PUR adhesives decreases with increasing WMC and temperature respectively.

The investigations mentioned above show that the adhesion of joints produced with adhesives containing fillers was noticeably increased. The goal of this study is to investigate if comparable improvements are also achievable for the use of 1C PUR adhesives under high temperature exposure.

3.4.3 Material and methods

Three laboratory adhesives were produced by Purbond (Sempach-Station, Switzerland) with a varying filler material content. Thereby chalk was mixed into the adhesive, using volume fractions of 15 and 30 %. The adhesives' parameters are listed in Table 3.4.1.

All bondings were carried out with beech wood (*Fagus sylvatica* L.). The raw density ρ at an equilibrium moisture content (EMC) ω of $(12 \pm 1)\,\%$ amounted to $(745 \pm 34)\,\mathrm{kg\,m^{-3}}$. The one-sided application of the adhesives was carried out with a spread of $150\,\mathrm{g\,m^{-2}}$ and a pressing pressure of $0.7\,\mathrm{MPa}$. To investigate the influence of the filler material content on the shear strength, 15 specimens of each group were tempered in a drying chamber for 1 h at 100 and 150°C respectively. Another group of specimens was conditioned at different relative humidities (35, 65, 85, 95 % RH) at a temperature of 20°C.

The shear strength was determined according to DIN EN 302-1 (2004). The specimens were tested using a displacement-controlled universal testing machine (Zwick Z100) under standard climatic conditions (20°C, 65 % RH). The shear strain ε was evaluated with a video-extensometer. After recording the stress-strain curve until failure, the wood failure percentage was estimated visually in steps of 10 %.

Table 3.4.2 Mean shear strength and median wood failure percentage of adhesive joints at varying climatic conditions

	Conditions		Adhesive A		Adhesive B		Adhesive C	
T [°C]	RH [%]	ω [%]	τ [MPa]	WF [%]	τ [MPa]	WF [%]	τ [MPa]	WF [%]
150	-	0	9.42 (1.30)	0	10.89 (3.28)	50	13.61 (2.28)	70
100	-	1.7	10.86 (2.18)	0	12.26 (2.41)	70	16.24 (2.43)	70
20	35	6.5	13.48 (1.85)	100	16.35 (1.47)	100	17.72 (1.90)	90
20	65	11.8	12.29 (1.85)	50	12.74 (2.38)	20	16.37 (1.54)	30
20	85	16.5	9.10 (2.02)	20	8.98 (2.49)	0	10.81 (1.43)	0
20	95	21.4	5.58 (2.70)	0	5.44 (1.87)	0	6.44 (2.54)	0

τ, mean tensile shear strength, standard derivation in brackets; WF, median wood failure percentage; T, temperature; RH, relative humidity; ω, mean wood moisture content

In addition we used an environmental scanning electron microscope (ESEM) and analysed the bond line by means of energy-dispersive X-ray spectroscopy (EDX) to investigate the penetration depth and distribution of the adhesives within the wood. The EDX analysis allows a chemical characterisation of the specimens and thereby to distinguish between adhesive, wood and filler material, which contains a high amount of calcium.

3.4.4 Results and discussion

The shear strength of the bonded wood joints increased significantly with a higher content of filler material. The graphs in Fig. 3.4.1(a) indicate an increase of strength at standard climatic conditions, but also after temperature exposure. The maximum increase amounted to 52 % at 100°C using 30 % filler. The wood failure percentage was also increased compared to adhesives without filler (Table 3.4.2) as a consequence of the better adhesion between wood and adhesive, which subsequently exceeded the wood strength.

The effect of the filler material decreased with increasing WMC. At 6.5 % wood moisture, the maximum overall increase of shear strength amounted to 31 % at 30 % filler material. Schrödter and Niemz (2006) determined maximum compression shear strength at about 12 % WMC within a similar investigation on commercially available 1C PUR adhesives. From this it follows that after the drying process internal compression stresses arise within the bond line which have a positive effect in the case of tensile load.

In contrast to the specimens exposed to high temperatures, the average increase in shear strength at 21.4 % WMC was relatively low (15 %); however, there was no significance at the 5 % level (Fig. 3.4.1(b)). This means that the filler material had no substantial effect on the shear strength at high WMC. The limiting factor for the adhesive bond is the moisture resistance of the adhesive itself, independent of its filler material content. Hydrolytic effects are a possible explanation for the lower shear strength.

3 Main investigations

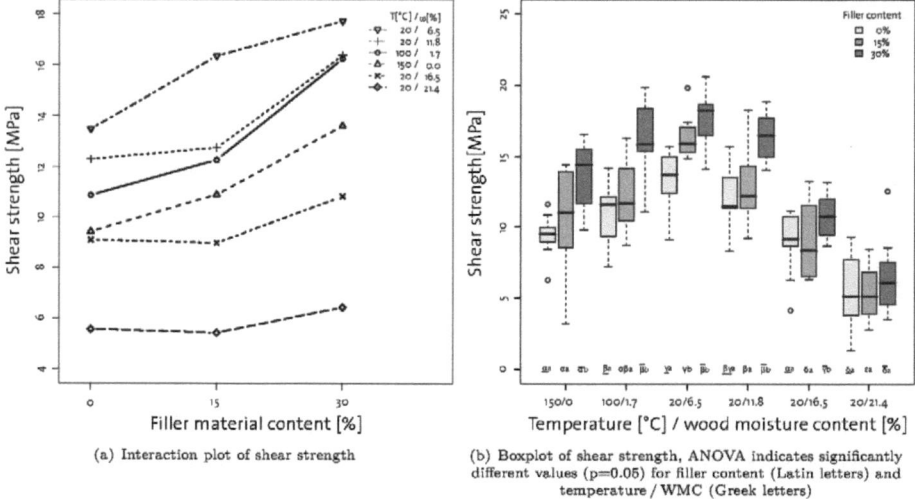

Fig. 3.4.1 Shear strength of 1C PUR adhesives depending on filler content and climatic conditions

The main reason for the increased shear strength is the reduced penetration into the cell lumina, which is clearly shown by the combined ESEM/EDX micrograph (Fig. 3.4.2). On the left side (30 % filler), a completely filled bond line and empty pores document a good bond. The adhesive without filler (right side) on the other hand, shows a poorly bonded adherend. The adhesive filled out pores even 500 µm away from the bond line, however the joint starved instead. Already Suchsland (1958) advised that there is no relationship between the penetration depth and the bonding quality, as long as the adhesive fills out the topmost surface forming cell layer.

Because calcium carbonate was used as filler material, the element calcium can be easily used for detecting the substance with EDX. It turned out that the filler material was homogeneously dispersed within the adhesive matrix (Fig. 3.4.2, picture detail) and no separations could be detected.

3.4.5 Conclusions

Chalk turned out to be a suitable filler material, which is easily addable to the adhesive, well miscible and cost efficient and it significantly improves the thermal stability of bonded wood joints in the aimed temperature range. For future studies it would be of particular interest to find suitable alternative filler materials and to determine the optimal filler material content, regarding costs and bonding properties.

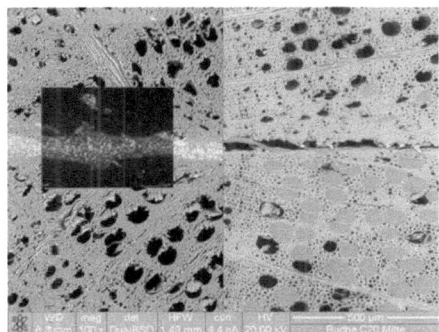

Fig. 3.4.2 ESEM micrograph of 1C PUR adhesive with 30 % filler (left), without filler (right) and EDX mapping of calcium (picture detail)

3.4.6 Acknowledgements

The authors would like to thank Gabriele Peschke from the Fracture Mechanics of Concrete Group at the Institute for Building Materials (ETH Zurich) for her support at the ESEM.

References

Beaud, F., Niemz, P. and Pizzi, A. (2006) Structure-property relationships in one-component polyurethane adhesives for wood: Sensitivity to low moisture content. J. Appl. Polym. Sci. 101 No. 6, 4181–4192

Dodiuk, H. et al. (2006) Polyurethane adhesives containing functionalized nanoclays. J. Adhes. Sci. Technol. 20 No. 12, 1345–1355

Kovačević, V. et al. (1996) Tensile properties of calcium carbonate-reinforced poly(vinyl acetate). J. Adhes. Sci. Technol. 10 No. 12, 1273–1285

Mansouri, H.R. and Pizzi, A. (2007) Recycled micronized polyurethane powders as active extenders of UF and PF wood panel adhesives. Holz Roh Werkst. 65 No. 4, 293–299

Muñoz-Milán, A.B. et al. (2005) Effect of the amount of calcium carbonate as filler on the rheological and adhesion properties of a water-based polyurethane dispersion. Macromol. Symp. 221 No. 4, 33–41

Park, S.W., Kim, B.C. and Lee, D.G. (2009) Tensile strength of joints bonded with a nanoparticle-reinforced adhesive. J. Adhes. Sci. Technol. 23 No. 19, 95–113

Richter, K. and Schierle, M. (2003) Behaviour of 1K PUR adhesives under increased moisture and temperature conditions. In **Teischinger, A.** and **Stingl, R.**, editors: Proceedings of the international Symposium on Wood Based Materials - Wood Composites and Chemistry. Vienna: Institute of Wood Science and Technology, BOKU Vienna, Lignovisionen 4, 149–154

Schrödter, A. and **Niemz, P. (2006)** Investigation on the failure behaviour of glue joints at high temperatures and relative humidity. Holztechnologie, 47 No. 1, 24–32

Sepulcre-Guilabert, J., **Ferrándiz-Gómez, T. P.** and **Martín-Martínez, J. M. (2001)** Properties of polyurethane adhesives containing natural calcium carbonate plus fumed silica mixtures. J. Adhes. Sci. Technol. 15 No. 2, 187–203

Suchsland, O. (1958) On the penetration of glue in wood gluing and the significance of the penetration depth for the strength of glue joints. Holz Roh Werkst. 16 No. 3, 101–108

Zeppenfeld, G. and **Grunwald, D. (2005)** Klebstoffe in der Holz- und Möbelindustrie. 2nd edition. Leinfelden-Echterdingen: DRW-Verlag, 368 p.

3.5 Paper V

J. Appl. Polym. Sci. (2012) 124:3641-3649

Influence of filler material on the thermal stability of 1C PUR adhesives

Sebastian Clauß[1], Dirk J. Dijkstra[2], Joseph Gabriel[3], Alexander Karbach[4], Mathias Matner[2], Walter Meckel[3], Peter Niemz[1]

[1]Institute for Building Materials, ETH Zurich, Schafmattstrasse 6, 8093 Zurich, Switzerland
[2]Bayer MaterialScience AG, Coatings, Adhesives & Specialties, Kaiser-Wilhelm-Allee 1, 51373 Leverkusen, Germany
[3]Purbond AG, Industriestrasse 17a, 6203 Sempach-Station, Switzerland
[4]CURRENTA GmbH & Co. OHG CHEMPARK, Solid State and Polymer Analysis, Rheinuferstrasse 7-9, 47829 Krefeld, Germany

3.5.1 Abstract

Filler materials are part and parcel of the adjustment of adhesives, in particular, their rheological and mechanical properties. Furthermore, the thermal stability of adhesives can be positively influenced by the addition of an expedient filler, with inorganic types common practice in most cases. In this study, one-component moisture-curing polyurethane adhesives for engineered wood products based on isocyanate prepolymers with different polymer-filled polyether polyols were investigated with regard to the filler's potential to increase the thermal stability of bonded wood joints. The property changes due to the addition of fillers were determined by means of mechanical tests on bonded wood joints and on pure adhesive films at different temperatures up to 200°C. Additional analyses by atomic force and environmental scanning electron microscopy advanced the understanding of the effects of the filler. The tested organic fillers, styrene acrylonitrile, a polyurea dispersion and polyamide, caused increases in the cohesive strength and stiffness over the whole temperature range. However, the selected filler type was hardly important with regard to the tensile shear strength of the bonded wood joints at high temperatures, although the tensile strength and Young's modulus of the adhesive films differed over a wide range. Prepolymers with a lower initial strength and stiffness resulted in worse cohesion, in particular, at high temperatures. This disadvantage, however, could be compensated by means of the filler material. Ultimately, the addition of filler material resulted in optimized adhesive properties only in a well-balanced combination with the prepolymer used.

3.5.2 Introduction

Adhesives in modern timber engineering must comply with a set of requirements to ensure accurate and safe processing as well as durable and hard-wearing use of the fabricated wood-construction elements. Given the wide range of requirements, one-component moisture-curing polyurethane (1C PUR) adhesives exhibit certain advantages because their properties can largely be controlled by a variation of the type and molar ratio of their components. The decisive parameters (in particular for thermal stability) are the free isocyanate (NCO) content and the cross-link density of the prepolymer (Na et al., 2005; Richter and Steiger, 2005; Richter, Pizzi and Despres, 2006; Šebenik and Krajnc, 2007; Chattopadhyay and Webster, 2009; Clauß et al., 2011b). The physical and chemical properties of the adhesives can be further adjusted by additives that are added subsequent to or during prepolymer synthesis (Clauß et al., 2011c). In addition to defoamers, plasticizers, organic solvents, wetting agents, dispersants, rheological agents, and catalysts, filler materials are also commonly used.

Filler materials are generally used for different purposes, with the substitution of a more expensive polymer being only one among many reasons. Furthermore, fillers can improve the mechanical properties, processability, thermal and dimensional stability, and fire retardancy of polymers (Xanthos, 2005). A multitude of different filler materials has been used to increase the mechanical properties and thermal stability of adhesives for timber engineering. Inorganic filler materials, such as calcite (Lučić, Kovačević and Hace, 1998), silica (Benli et al., 1998; Lučić, Kovačević and Hace, 1998; Sepulcre-Guilabert, Ferrándiz-Gómez and Martín-Martínez, 2001), kaolin (Lučić, Kovačević and Hace, 1998), calcium carbonate (Kovačević et al., 1996; Sepulcre-Guilabert, Ferrándiz-Gómez and Martín-Martínez, 2001; Muñoz-Milán et al., 2005), chalk (Clauß et al., 2011a), carbon black (Benli et al., 1998; Park, Kim and Lee, 2009), nanoclays (Benli et al., 1998; Lučić, Kovačević and Hace, 1998; Dodiuk et al., 2006), aluminum oxide (Benli et al., 1998), and zirconium(III) oxide (Benli et al., 1998) have, therefore, been investigated in combination with different types of adhesives, such as urea-formaldehyde and phenol–formaldehyde (Mansouri and Pizzi, 2007), poly(vinyl acetate) (Kovačević et al., 1996; Lučić, Kovačević and Hace, 1998), 1C PUR (Clauß et al., 2011a), solvent-borne PUR (Sepulcre-Guilabert, Ferrándiz-Gómez and Martín-Martínez, 2001), polyurethane dispersions (Muñoz-Milán et al., 2005), thermosetting PUR (Dodiuk et al., 2006), and epoxy (Benli et al., 1998; Park, Kim and Lee, 2009), but organic filler types, such as PUR powder (Mansouri and Pizzi, 2007) and polyurea dispersions (PHDs) (Spitler and Lindsey, 1981), are also described in the literature. The thermomechanical properties can be modified by reinforcement of the polymer matrix with expedient fillers. The higher stiffness and strength of the filler are, therefore, used to improve the overall properties of the composite. The improvement of the thermal stability depends on the type, shape, size, and amount of the filler material used. The uniformity of the dispersion is of particular importance and is a disqualifying criterion for many types of fillers because the storage stability over several months must be guaranteed. Furthermore, the aspect ratio and the degree of interaction between the inorganic fillers and the polymer matrix are vitally important (Lučić, Kovačević and Hace, 1998; Xanthos, 2005). Most suitable are

Table 3.5.1 Chemical and physical properties of the prepolymers and adhesives

ID-code	P1	P1SAN5	P1SAN7	P1PHD5	P1PHD7	P1PA5	P2	P2PHD5	P2PHD10
c_{NCO} [%]	16	15.5	15.5	15.5	15.5	15.5	14.3	14.3	14.3
f	2.8	2.8	2.8	2.8	2.8	2.8	2.4	2.4	2.4
CLD [mol kg^{-1}]	1.09	1.09	1.09	1.09	1.09	1.09	0.55	0.55	0.55
c_{UG} [mol kg^{-1}]	0.6	0.6	0.6	0.6	0.6	0.6	0.26	0.26	0.26
η [mPas]	17000	20300	31400	23450	31800	24500	25900	59000	67200
t_{open} [min]	80	84	84.5	82.5	83	81	80	68	67.5
Filler	-	SAN	SAN	PHD	PHD	PA	-	PHD	PHD
c_{filler} [%]	-	5	7.5	5	7.5	5	-	5	10
Defoamer	yes	yes	yes	yes	yes	yes	yes	yes	yes
Pyrog. silica	-	-	-	-	-	yes	yes	yes	yes
Catalyst	-	-	-	-	-	-	yes	yes	yes

c_{NCO}, NCO content; f, functionality; CLD, cross-link density; c_{UG}, urethane group content; η, viscosity at 20°C; t_{open}, open time; c_{filler}, filler material content; P, prepolymer; PHD, polyurea dispersion; SAN, styrene acrylonitrile; PA, polyamide

fillers of nanoscale dimensions that are uniformly dispersed and interact strongly with the organic matrix (Chattopadhyay and Webster, 2009).

This investigation was focused on organic filler materials with the potential to improve the thermal stability of 1C PUR adhesives for engineered wood products. Polymer-filled polyether polyols, therefore, came into consideration on the basis of their several advantages over inorganic fillers:

1. Because of the similar density of organic filler materials compared to NCO-terminated prepolymers, filled systems have a better storage stability, even at lower viscosities.

2. Finely dispersed organic particles can increase the interaction by secondary forces, mainly hydrogen bonds, between the filler and the urethane and urea groups of the polyurethane matrix.

The aim of this study was to investigate whether the advantages of organic fillers also have an effect on the thermal stability of 1C PUR adhesives. Different variants of filled prepolymers were, therefore, produced, formulated, and subsequently tested with regard to their cohesive strength by means of tensile tests on adhesive films and with regard to their bonding performance by means of tensile shear strength tests on bonded wood joints at elevated temperatures.

3.5.3 Material and methods

Prepolymers and adhesives

The adhesives investigated in this study (Table 3.5.1) were based on two different prepolymers produced by Bayer MaterialScience (Leverkusen, Germany). Mixtures of methylene diphenyl diisocyanate (MDI) isomers consisting mainly of 4,4'- and 2,4'-MDI and polymer MDI with a functionality greater than 2 were used. The NCO content amounted to 16 % for prepolymer 1 (P1) and 14.3 % for prepolymer 2 (P2). The estimated functionalities totaled 2.8 and 2.4 for P1 and P2, respectively.

3 Main investigations

Three types of filler material, styrene acrylonitrile (SAN), PHD, and polyamide (PA) powder, were investigated in this study. PHD is the branch copolymer product of the polyaddition reaction of a polyisocyanate, a polyamine, and a polyether polyol (König and Dietrich, 1978). When used as a filler material, polyurea particles are finely dispersed in the polyether polyol. The PHD polyols used contained 5 % (P1PHD5) and 7.5 % (P1PHD7) PHD in combination with P1 and 5 % (P2PHD5) and 10 % (P2PHD10) PHD solid content in combination with P2. The filler appears as an opaque white dispersion with a median particle diameter of less than 1 µm (Ionescu, 2005).

SAN is a random amorphous copolymer of styrene and acrylonitrile monomers that has improved mechanical properties and better chemical resistance than polystyrene (Alberts and Ballé, 1982). The used SAN contained about 40 % acrylonitrile. Non-cross-linking, free-flowing PA powder characteristically has good mechanical properties, even at elevated temperatures, and good chemical resistance. The used powder had a median particle size of 10 µm and a specific gravity of $1\ \mathrm{g\,cm^{-3}}$.

In the case of SAN- and PHD-filled polyols, the fillers were produced and dispersed in a standard polyether by Bayer MaterialScience and, depending on the formulation, mixed with additional polyol components and subsequently reacted with NCO during the prepolymerization process. The appropriate products were provided by Bayer MaterialScience. In contrast to the previous fillers, PA powder was added during the subsequent formulation of the adhesive by Purbond (Sempach-Station, Switzerland). The powder was dispersed in the prepolymer, together with defoamers, rheology modifiers, and catalysts. These additives are commonly used to prepare standard commercial adhesives. An amine catalyst was used to set a similar open time between 60 and 90 min for all of the formulated adhesives.

Tensile test on adhesive films

The adhesive films were produced by application of the liquid adhesive on a plastic sheet; we ensured a constant application thickness with a special applicator. We minimized the typical foaming effect of polyurethanes by applying a film thickness of roughly 0.25 mm and performing the reaction under 50 % relative humidity (RH). Once the reaction had progressed sufficiently, the films were removed from the sheet and stored for a minimum of 3 days under standard climatic conditions (20°C, 65 % RH). The actual samples were punched with sample shape type 1B according to DIN EN ISO 527-3 (2003).

The tensile properties of the films were obtained according to DIN EN ISO 527-1 (1996). In addition to standard climatic conditions, tests also were performed at 70 and 200°C. The measurements were carried out with a Zwick Z100 universal testing machine with a 500 N load cell with a testing speed of $2\ \mathrm{mm\,min^{-1}}$ until failure. Tensile and transverse deformation was recorded optically with a Messphysik (Fürstenfeld, Austria) video-extensometer ME-46. Young's modulus, tensile strength, and strain at maximal load and Poisson's ratio were determined from the load-displacement curves. The mean values were calculated for a series of at least six specimens.

Dynamic mechanical analysis (DMA)

The films for the DMA specimens were prepared as described previously. Samples 20 mm in length and 4 mm in width were cut from them and evaluated in a Seiko (Chiba, Japan) DMS 210 apparatus in the tensile mode over a temperature range of -140 to 250°C at a frequency of 1 Hz and at a ramp rate of $2°C\,min^{-1}$. The underlying standard for this test was ISO 6721-4 (1994).

Differential scanning calorimetry (DSC)

The thermal properties of the polyurethane prepolymer films were analyzed with a PerkinElmer (Waltham, MA) DSC-7 differential scanning calorimeter. Approximately 10 mg of polyurethane film was placed in a standardized pan with caps at a heating rate of $20°C\,min^{-1}$. Two consecutive runs were carried out, with heating from -100 to 100°C followed by cooling down to -100°C (cooling rate = $320°C\,min^{-1}$) and nitrogen flushing before the second heating run from -100 to 100°C. The glass-transition temperatures (T_g's) were determined at half height of the glass step.

Viscometry

The viscosity of the prepolymers was determined at 23°C with an Anton Paar (Graz, Austria) MCR 301 cone/plate rheometer (d (cone diameter) = 25 mm, α (cone angle) = 1°) at a shear rate of $150\,s^{-1}$ according to DIN 53019 (2008).

Tensile shear test on bonded wood joints

The longitudinal tensile shear strength of bonded wood joints was tested according to DIN EN 302-1 (2004). The specimens were prepared from beech (*Fagus sylvatica* L.), which is characterized by a low content of extractives (to avoid chemical interactions with the adhesives) and a high strength compared to spruce, which is commonly used in timber engineering in Europe. The raw density of the adherend was $745 \pm 50\,kg\,m^{-3}$ at an equilibrium moisture content of about $12 \pm 1\,\%$. The adhesives were applied to one side by means of a toothed spatula with a spread of $200\,g\,m^{-2}$. The pressing time was at least 3 h at a pressure of about 0.8 MPa.

The shear tests were likewise performed at 20, 70, and 200°C. The specimens were randomized to ensure that the wood and processing effects did not introduce bias in the estimation of the tensile shear strength at different temperatures. Sample groups of 15 specimens were tempered at the same time in a drying chamber for 1 h before testing.

The test was displacement-controlled with a universal testing machine (Zwick/Roell (Ulm, Germany) Z010). The strain was measured by means of a clip-on displacement transducer. After failure of the specimen, the wood failure percentage (WFP) was estimated visually in steps of 10 %.

3 Main investigations

Table 3.5.2 Results of the tensile test of films and bonded wood joints

	T [°C]	P1	P1SAN5	P1SAN7	P1PHD5	P1PHD7	P1PA5	P2	P2PHD5	P2PHD10
τ [MPa]	20	11.06	10.63	9.18	10.23	11.19	13.15	10.87	13.21	13.23
	70	11.28	10.7	11.43	10.46	13.23	13.25	9.79	11.61	12.86
	200	8.41	9.88	10.9	9.71	10.39	10.62	6.96	8.95	10.76
WFP [%]	20	0	0	0	0	20	30	0	20	20
	70	10	0	0	0	50	40	0	0	10
	200	0	0	60	0	100	90	0	0	40
σ [MPa]	20	26.5	37.6	35.5	39.3	39.5	35.1	14	16.1	25.0
	70	37.1	47.5	48.1	47.3	49.2	38.1	16	19.2	25.4
	200	29	37.8	38.9	38.4	41	30.3	13.8	17.9	21.0
E [MPa]	20	1360	1965	1899	2016	2019	1716	449	616	945
	70	1199	1646	1867	1763	1823	1251	356	503	798
	200	478	1105	1009	985	1051	464	182	244	387

τ, tensile shear strength; WFP, wood failure percentage; σ, tensile strength; E, Young's modulus; T, temperature; P, prepolymer; PHD, polyurea dispersion; SAN, styrene acrylonitrile; PA, polyamide

Microscopic analysis

For a qualitative analysis of the bond line and interphase region, micrographs were taken with a dualbeam scanning electron microscope (FEI (Hillsboro, OR) Quanta 200 3D) in low-vacuum mode. Atomic force microscopy (AFM) imaging was performed in tapping mode (Digital Instruments (Tonawanda, NY) D3000) to provide phase-contrast images of the polymer structure.

3.5.4 Results and discussion

The investigated film and bonding parameters (Table 3.5.2) of the cured polyurethane prepolymers and modified adhesives were affected differently by the temperature, prepolymer configuration, and type and the amount of filler material. Furthermore, factorial analysis detected significant interactions between these individual influencing factors. In the following text, the temperature-dependent material behavior of the prepolymers and the formulated adhesives based on them are discussed with consideration of the adhesive parameters varied. The influence of the temperature was clearly revealed by the Young's modulus (E), which was affected to a much higher degree by the heat treatment than the tensile strength (σ) of the films or the tensile shear strength (τ) of the bonds. Ultimately, however, the result of the bonded wood assemblies is determining for the use of an adhesive for structural wood bonding.

Influence of the prepolymer configuration

The prepolymer configuration had a clear impact on the temperature-dependent material behavior of the pure adhesive films and the bonded wood joints. Because the material behavior was prescribed primarily by the prepolymer configuration (Clauß et al., 2011b), all tested adhesives revealed linearly decreasing moduli in the observed temperature range. The comparison of the prepolymers revealed that σ and E of P1 at 20°C were about 50 and 30 % higher, respectively, compared with the values of P2, with these percentages increasing with rising temperature. The use of branched polyfunctional components ensured a high density of covalent cross-linking with high thermal stability. The cross-linking of the hard segments was supported by hydrogen bonding of the NH groups and carbonyl groups of urea and urethane linkages. As already shown by Clauß et al. (2011b), the higher hard-segment content and higher crosslink density were responsible for the stronger polymer network, which is why P1 was found to exhibit better thermal stability with rising temperatures. At higher temperatures, the polymer chain flexibility increased partly because of the disappearance of the secondary hydrogen bonds in the polymer structure. The higher flexibility caused a more ductile material behavior and, thus, a higher strain to failure at a somewhat lower maximal strength. This was confirmed by the stress–strain curves in Fig. 3.5.1(a) compared with those in Fig. 3.5.1(c), which show a more ductile behavior and a larger strain at break for P1 and an almost unchanged mechanical behavior for P2.

The comparison of the complex moduli obtained by DMA measurements (Fig. 3.5.2) confirmed again significant differences in the temperature-dependent performance of the cured prepolymers. P2 exhibited a very distinct glass transition at about -62°C, which was related to the polyether component. Starting from this temperature, the storage modulus (E') of P2 decreased drastically from about $E' = 4750$ to $1130\,\text{MPa}$ at 20°C. The curve of P1, on the other hand, showed a slow and gradual decrease of E' with $E' = 1600\,\text{MPa}$ at 20°C. Several small maxima in the loss modulus of sample P1, an indication for a very complicated polymer blend morphology, could be seen. At about 63°C, P1 exhibited a somewhat larger transition peak in the loss modulus, probably caused by the T_g of the hard segment. The curves of both prepolymers merged gradually at higher temperatures, up to about 150°C, where they were nearly equal to each other.

Although σ remained almost constant from 20 to 70°C, the ductility of the polymer films increased significantly. At much higher temperatures, σ decreased to the base level at 200°C (Fig. 3.5.1(a), (c), (e)). The maximum in σ of sample P1 at 70°C was probably caused by the glass transition in that range. Measurements by DSC (not shown) revealed a caloric effect in the temperature range around 70°C, which confirmed the glass transition found in DMA. However the second heating in DSC measurements did not show a glass-transition temperature, which might have been caused by a change in morphology during the first heating. It was not possible to give a definite explanation at the moment, but investigations into this issue are still in progress.

At 200°C, E decreased by about 60 % compared with the values obtained at 20°C (Fig. 3.5.1(a) and (e)). σ, by comparison, reached values in the range of the values obtained at 20°C. The stress–strain curves showed, for all adhesives, an elastic behavior without a yield point and maximal strains of more than 10 %. The differentiability of the adhesives was similar to the results obtained at 20°C. The effect of the temperature, however, had to be interpreted cautiously because the testing itself was carried out at 20°C after heating. Although the samples were taken rapidly from the oven to the testing machine, the temperature decreased significantly because of the low sample mass.

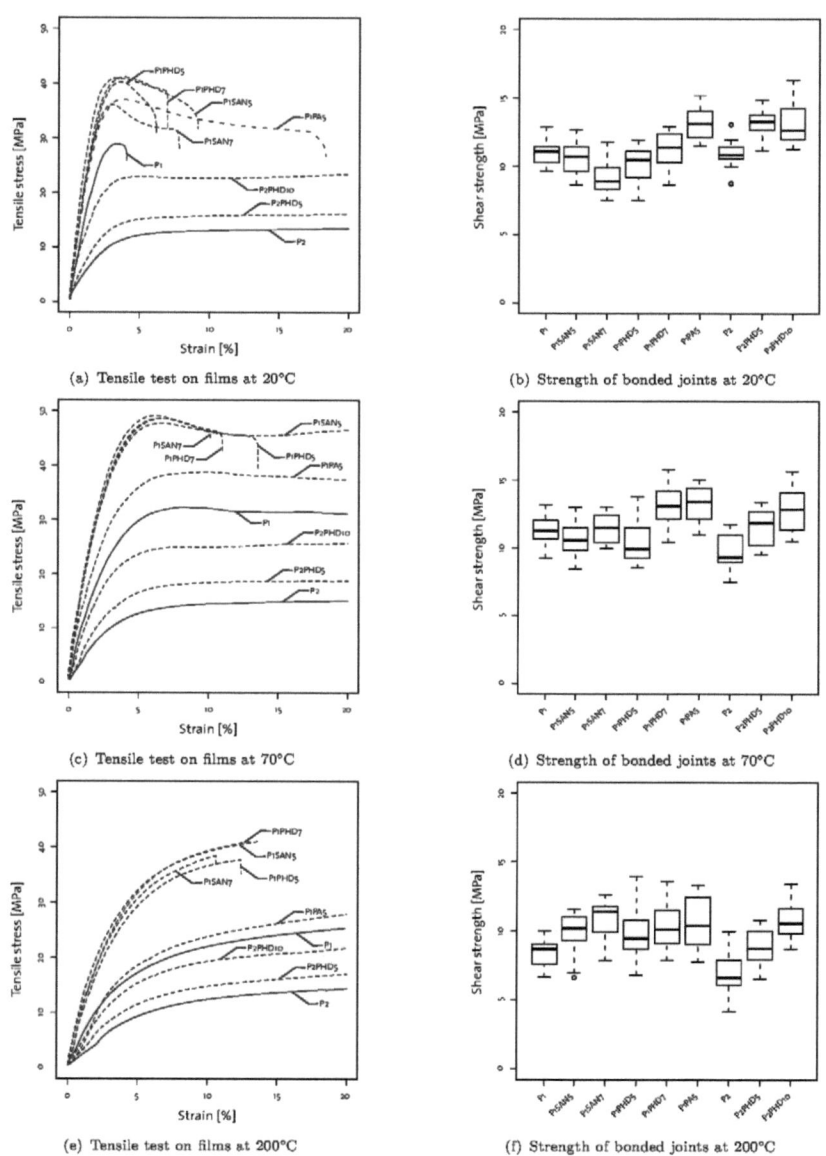

Fig. 3.5.1 Stress-strain curves of films from prepolymers and adhesives filled with PHD, SAN and PA tested in tension, and tensile shear strength of bonded wood joints

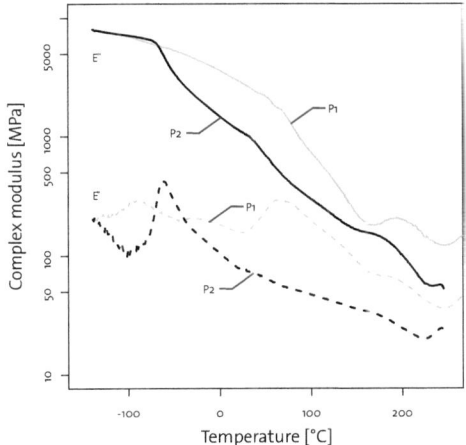

Fig. 3.5.2 Storage and loss modulus of prepolymers with different hard segment content and cross-link density

In contrast to the results for the prepolymer films, the results for the bonded joints show no differentiation between the two prepolymers at room temperature (Fig. 3.5.1(b)). Because WFP (Table 3.5.2) was on a comparatively low level, we assumed that the prepolymers failed and not the adherend. The macroscopic assessment of the fracture surface indicated an adhesive failure of the bond. In the majority, the failure occurred because of an insufficient adhesion between the adhesive and substrate, which is clearly shown in Fig. 3.5.3(c) and (d). The use of a UV light source revealed that the adhesive was located at one of the involved adherends. A cohesion failure was, therefore, out of the question.

At 70°C, the same differentiation as found for the film properties was found, with P1 exhibiting a significantly higher τ than P2 (Fig. 3.5.1(d)). The increase in the temperature to 200°C resulted in decreases in τ by about 30 % for P1 and 55 % for P2. Finally, the bonded joints also reflected the differences in the prepolymer composition, albeit not at 20°C.

Influence of the filler material type

Three types of filler materials were used with the intention of increasing the thermal stability of the prepolymers. A comparison of the adhesives with 5 % filler content showed that SAN and PHD exhibited nearly the same results for the film and bonding properties. Both adhesives achieved significantly higher strengths and stiffnesses compared with the unfilled prepolymer at all of the tested temperatures. The tensile shear strength, however, was positively affected only at 200°C (Fig. 3.5.1(f)).

Compared with P1, P1PA5 also exhibited increased values for E and σ, but the values were much closer to those of the prepolymer compared to P1SAN5 or P1PHD5, in particular, at 70 and 200°C.

3 Main investigations

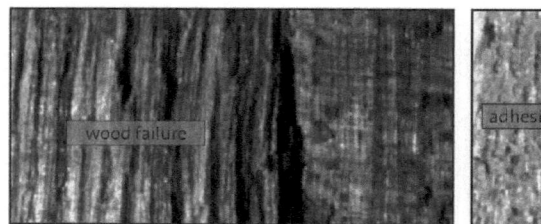
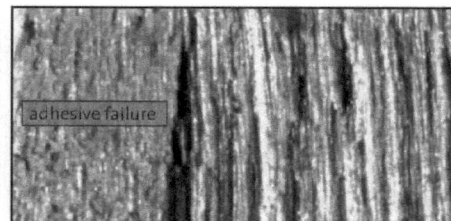

(a) Top side of fracture surface with visible light source

(b) Bottom side of fracture surface with visible light source

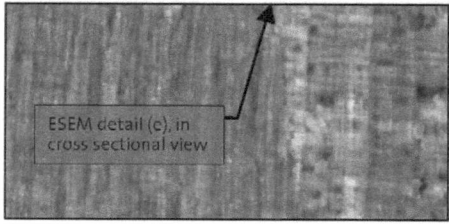
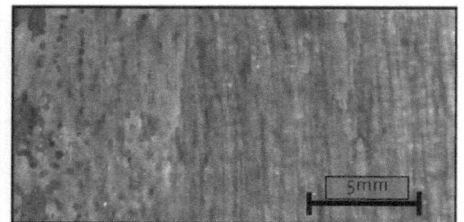

(c) Top side of fracture surface with UV light source

(d) Bottom side of fracture surface with UV light source

Fig. 3.5.3 Stereo light micrographs of a bonded wood joint after tensile shear testing

The stress–strain curves of P1 and P1PA5 revealed a more elastic behavior at 70°C, whereas the adhesives filled with SAN and PHD exhibited a rather ductile behavior. The tensile shear strength and WFP, however, significantly increased for P1PA5 at all of the tested temperatures.

The results led us to the conclusion that an improvement of the cohesion did not necessarily cause an improvement in the bonding performance. The higher stiffness of the adhesives filled with PHD or SAN may have possibly been disadvantageous with respect to the bonding strength at room temperature. Because the reactivity and viscosity of the adhesives with 5 % filler material ranged on the same level (Table 3.5.1), the influence of these factors could virtually be neglected.

The AFM phase-contrast images (Fig. 3.5.4) showed that the fillers differentiated clearly from the prepolymer structure, which revealed no obvious difference between the pure prepolymer (Fig. 3.5.4(a)) and the formulated prepolymer without filler material (Fig. 3.5.4(b)). The presence of a phase-segregated system became visible as bright, hard-segment-rich phases and dark, soft-segment-rich phases within the polymer structure. The filler materials, PHD (Fig. 3.5.4(c)) and SAN (Fig. 3.5.4(d)), differed significantly with respect to their size and dispersion. PHD exhibited larger and agglomerated structures, whereas SAN was homogeneously dispersed, with particle diameters up to 100 nm.

Considering the interaction between the prepolymer type and the filler material, it can be stated that the presence of PHD caused, in the case of P1, a significant increase in the film properties at all tested

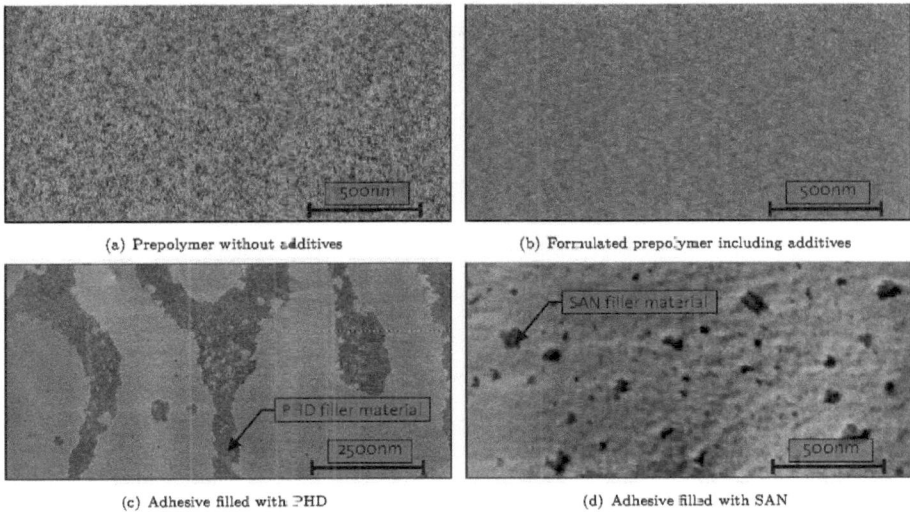

Fig. 3.5.4 Atomic force microscopy (AFM) phase images

temperatures but, aside from 200°C, no significant difference in τ. The combination of P2 with the same filler, however, revealed moderate increases in E and σ but a large increase in τ of up to 30 % at 200°C. Surprisingly, P1PHD5 also exhibited an increase in τ at this temperature of about 15 % so that, in the end, this combination achieved the best thermal stability of the bondings as a result of the higher initial strength of prepolymer P1.

Influence of the filler material content

The effect of the filler material content on the film properties was completely different for the two prepolymers investigated. In case of P1, SAN and PHD resulted in almost similar values of E between 1.9 and 2.0 GPa and σ's between 37.6 and 39.5 MPa with both contents, with the only significant differences being in σ ($\sigma = 35.5$ MPa) in the case of P1SAN7. P2, in combination with PHD, however, revealed statistically significant increases in σ and E of about 25 and 35 %, respectively. Because the difference between the two filler material contents was only 2.5 % in the case of P1, it may be possible that an influence was not detected for this reason in view of the statistical spread.

In the case of P1, statistically significant effects of the filler material content on τ were shown at 20 and 70°C. In the case of P2, an increase in the filler material content from 5 to 10 % PHD revealed significant effects at 70 and 200°C. Increasing the filler material led, in both cases, to an increase in τ of about 15 %, and the higher filler material content was further accompanied by an increase in

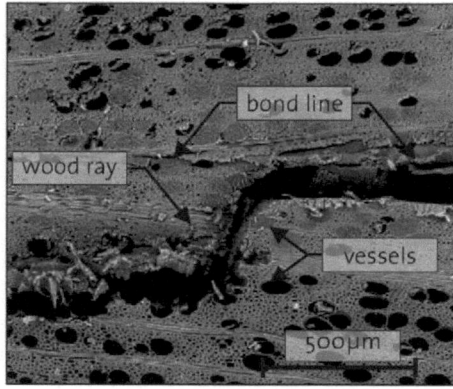

Fig. 3.5.5 ESEM micrograph (magnification = 100x) of the cross section showing a change in the crack propagation from adherend (wood failure) into the bond line (adhesive failure)

the WFP. One reason for this behavior could be found in a delayed penetration and, consequently, a higher concentration of the adhesive in the bond line (Clauß et al., 2011a). Another aspect could have been the fact that liquid prepolymer penetrated into the wood and resulted in a bond line with a higher filler content, which increased the modulus.

SAN showed no improvement of the bonding when the filler material content was increased from 5 to 7.5 %. A significant difference between the different contents was found only at 20°C. Surprisingly, τ decreased in this case.

Influence of the substrate

Because WFP increased with increasing temperature, the drop in bonding strength was caused, on the one hand, by the decreasing polymer strength and, on the other hand, by the wooden adherend. The stereo light micrographs in Fig. 3.5.3(a) and (b) show the two fracture surfaces of the same sample with about 60 % wood failure. The crack followed the lowest resistance, which partly occurred within the substrate and partly between the substrate and adhesive (Fig. 3.5.5). Wood rays and areas with changing wood density strongly influenced the fracture behavior of the individual samples.

The resulting wood failure at higher temperatures was the consequence of greater substrate degradation compared with the adhesive. An increasing temperature at a constant or decreasing RH caused a decrease in the wood moisture content due to the hygroscopic behavior of the material. Besides drying stresses, additional internal stresses developed as a consequence of the fixation of the adherends by the adhesive. The drying process, furthermore, reduced the elasticity and tensile shear strength of the wood. A linear correlation between τ and the temperature of different wood species was shown in a previous study by Clauß et al. (2011b).

The reasons for the strength reduction from about 65°C are depolymerization reactions within the wood structure that occur without significant weight loss. The chemical bonds of wood start breaking at temperatures higher than 100°C (White and Dietenberger, 2001). A pyrolysis of the wood components was not to be expected in the temperature range we investigated. Consequently, we assumed that all filled adhesives based on P1 and P2 filled with 7.5 and 10 % PHD, respectively, PHD exhibited higher τ than the adherend. Values obtained for beech under similar conditions lay at about 8.42 MPa (Clauß et al., 2011b), compared with values over 9.5 MPa obtained for the adhesives mentioned.

3.5.5 Conclusions

- Adhesives based on prepolymers with higher initial strengths and stiffnesses reveal better cohesion over the entire temperature range, which was why the filler material had a greater impact for more elastic polymers.
- The addition of the filler material caused an increase in the cohesive strength and stiffness over the entire temperature range. However, no substantial difference between PHD and SAN was found.
- PA powder revealed a lower strength and stiffness but better bonding performance at low temperatures compared with SAN or PHD.
- At high temperatures (200°C), the type of filler material was hardly important with respect τ of the bonded wood joints, although σ and E of the adhesive films differed over a wide range.
- The filler materials ensured an adequate bond line between the adherends, although because of the porosity of the substrate, low-viscosity adhesive fractions tended to penetrate into it. If penetration occurred, the stiffness of the bond line was increased by the relative increase in filler material content; this could be considered as an advantage.
- Organic filler materials have, compared to, for example, chalk, the special advantage of a very similar specific density, which results in improved storage stability for even low-viscosity adhesives because segregation is virtually precluded.
- PHD fillers ensure very good cohesion to an adhesive matrix because of hydrogen bonding.

References

Alberts, H. and **Ballé, G.** (1982) Verfahren zur Herstellung von modlflzlerten Polyiitherpolyolen und deren Verwendung zur Herstellung von Polyurethankunststoffen., European Patent, No. 0008444B1

Benli, S. et al. (1998) Effect of fillers on thermal and mechanical properties of polyurethane elastomer. J. Appl. Polym. Sci. 68 No. 7, 1057–1065

Chattopadhyay, D.K. and Webster, D.C. (2009) Thermal stability and flame retardancy of polyurethanes. Prog. Polym. Sci. 34 No. 10, 1068–1133

Clauß, S. et al. (2011a) Improving the thermal stability of one-component polyurethane adhesives by adding filler material. Wood Sci. Technol. 45 No. 2, 383–388

Clauß, S. et al. (2011b) Influence of the chemical structure of PUR prepolymers on thermal stability. Int. J. Adhes. Adhes. 31 No. 6, 513–523

Clauß, S. et al. (2011c) Influence of the adhesive formulation on the mechanical properties and bonding performance of polyurethane prepolymers. Holzforschung, 65 No. 6, 835–844

DIN EN 302-1 (2004) Adhesives for load-bearing timber structures – Test methods – Part 1: Determination of bond strength in longitudinal tensile shear strength. Berlin: Beuth Verlag

DIN EN ISO 527-1 (1996) Plastics – Determination of tensile properties – Part 1: General principles. Berlin: Beuth Verlag

DIN EN ISO 527-3 (2003) Plastics – Determination of tensile properties – Part 3: Test conditions for films and sheets. Berlin: Beuth Verlag

Dodiuk, H. et al. (2006) Polyurethane adhesives containing functionalized nanoclays. J. Adhes. Sci. Technol. 20 No. 12, 1345–1355

Ionescu, M. (2005) Chemistry and technology of polyols for polyurethanes. Shrewsbury: Smithers Rapra Press, 585 p.

ISO 6721-4 (1994) Plastics – Determination of dynamic mechanical properties – Part 4: Tensile vibration - Non-resonance method. Berlin: Beuth Verlag

König, K. and Dietrich, M. (1978) Stable polyurethane dispersions and process for production thereof., United States Patent, No. 4,089,835

Kovačević, V. et al. (1996) Tensile properties of calcium carbonate-reinforced poly(vinyl acetate). J. Adhes. Sci. Technol. 10 No. 12, 1273–1285

Lučić, S., Kovačević, V. and Hace, D. (1998) Mechanical properties of adhesive thin films. Int. J. Adhes. Adhes. 18 No. 2, 115–123

Mansouri, H.R. and Pizzi, A. (2007) Recycled micronized polyurethane powders as active extenders of UF and PF wood panel adhesives. Holz Roh Werkst. 65 No. 4, 293–299

Muñoz-Milán, A.B. et al. (2005) Effect of the amount of calcium carbonate as filler on the rheological and adhesion properties of a water-based polyurethane dispersion. Macromol. Symp. 221 No. 1, 33–41

Na, B. et al. (2005) One-component polyurethane adhesives for green wood gluing: Structure and temperature-dependent creep. J. Appl. Polym. Sci. 96 No. 4, 1231–1243

Park, S.W., Kim, B.C. and Lee, D.G. (**2009**) Tensile strength of joints bonded with a nanoparticle-reinforced adhesive. J. Adhes. Sci. Technol. 23 No. 19, 95–113

Richter, K., Pizzi, A. and Despres, A. (**2006**) Thermal stability of structural one-component polyurethane adhesives for wood - Structure-property relationship. J. Appl. Polym. Sci. 102 No. 1, 24–32

Richter, K. and Steiger, R. (**2005**) Thermal stability of wood-wood and wood-FRP bonding with polyurethane and epoxy adhesives. Adv. Eng. Mater. 7 No. 5, 419–426

Šebenik, U. and Krajnc, M. (**2007**) Influence of the soft segment length and content on the synthesis and properties of isocyanate-terminated urethane prepolymers. Int. J. Adhes. Adhes. 27 No. 7, 527–535

Sepulcre-Guilabert, J., Ferrándiz-Gómez, T.P. and Martín-Martínez, J.M. (**2001**) Properties of polyurethane adhesives containing natural calcium carbonate plus fumed silica mixtures. J. Adhes. Sci. Technol. 15 No. 2, 187–203

Spitler, K.G. and Lindsey, J.J. (**1981**) Phd polyols, a new class of pur raw-materials. J. Cell. Plast. 17 No. 1, 43–50

White, R.H. and Dietenberger, M.A.; Buschow, K.H.J. et al., editors (**2001**) Chap. Wood Products: Thermal degradation and fire. In Encyclopedia of materials: Science and technology. 2nd edition. Oxford: Elsevier Science Ltd., 9712–9716

Xanthos, M. (**2005**) Functional fillers for plastics. Weinheim: Wiley-VCH, 432 p.

4 Synthesis

4.1 Main findings

The main findings obtained in this thesis can be subdivided according to the adhesive development from the prepolymer to the formulated adhesive and finally to the filled adhesive.

Prepolymers It could been shown that the thermal stability is mainly influenced by the urea group content and the cross-link density of the reacted system. The tensile shear strength of the bonded wood joints as well as the stiffness of prepolymer films is significantly increased by a higher content of urea hard segments at all tested temperatures. Consequently, properties of the polymer are all the more dominated by the hard segment domain the higher the content of free NCO groups of the prepolymer is. In general the polyisocyanate component is mostly responsible for the properties (elasticity, cohesion and strength) of the finished product, which allows adjustment of the prepolymer properties by varying the molar ratio of the components. Urethane hard segments have only a minor effect on the measured properties. Below 100°C it seems that a higher urethane content contributes positively to the tensile shear strength of bonded joints as well as to the storage modulus of films. Above 180°C however, the increased urethane content has rather negative effects since the thermal stability of these groups is significantly lower compared to urea hard segments.

Ethylene oxide glycol ethers gain no improvement in the thermal stability. Ether mixtures consisting only of PO ethers reveal significantly higher stiffness after passing the glass step in the range of about 50 to 100°C. Branched cross-linking is attainable either via the polyether or via the isocyanate component. The obtained prepolymers revealed the same behavior regarding thermal stability. A too low cross-linking of the polyurethane prepolymer can cause, in both cases, segregation into hard and soft segment-rich phases of the polymer with differing mechanical properties, which can negatively affect the bonding performance.

Formulated adhesives The comparison between prepolymers and formulated adhesives based on them showed that the mechanical properties of the 1C PUR adhesives are completely determined by the prepolymer composition. The formulation of the adhesive by the addition of defoamer, plasticizer, organic solvents, wetting agents, dispersants, rheological agents and catalysts (with the exception of fillers) revealed almost no influence on the cohesion. However, by adjusting the viscosity and reactivity with the addition of these additives, the bonding performance of formulated adhesives is significantly enhanced in comparison with unformulated prepolymers. Possible negative effects due to side reactions that may occur by the addition of catalyst or wetting agents can be excluded. Also segregations of pyrogenic silica, which can compromise the bonding performance, could not be found.

Compared to additionally tested amino- and phenoplastic resins, 1C PUR adhesives are characterized by significantly lower stiffness and hardness, but they are able to absorb much more deformation energy and show ductile failure behavior leading to lower wood failure. Partly an optimal prepolymer configuration regarding the thermal stability (high cross-linking and urea hard segment content) jars with acceptable practical processing conditions (viscosity, foaming, etc.). By means of the formulation it is possible to counter these effects, however, it is the overall goal to find a good compromise between the mechanical properties of the finished product and its rheological and kinetic properties.

Filled adhesives Filler materials are well suited to enhance the thermal stability of 1C PUR adhesives. Especially adhesives based on prepolymers with lower initial strength and stiffness reveal better cohesion over the entire temperature range after the addition of filler materials. Due to the high initial strength and stiffness of the highly cross-linked prepolymers in this case, causes the fillers to have a lighter impact on the cohesion. In principle all tested types, regardless of whether they are inorganic or organic, significantly improved the results compared to unfilled adhesives. However, at high temperatures (200°C), the type of filler material reduces in significance with respect to the tensile shear strength of bonded wood joints, although the tensile strength and Young's modulus of the adhesive films differ over a wide range. No substantial differences were found between the organic prepolymer-incorporated fillers PHD and SAN.

The additionally incorporated PA powder reveals lower tensile strength and stiffness, but better bonding performance at low temperatures compared with SAN or PHD. Inorganic fillers, such as chalk, are very cost efficient compared to organic fillers, they are easily addable to the adhesive and well miscible and they also significantly improve the thermal stability of bonded wood joints. Organic filler materials have, compared to chalk, the special advantage of a very similar specific density, which results in improved storage stability for even low-viscosity adhesives since segregation is virtually precluded. Due to hydrogen bonding, suited organic fillers ensure very good cohesion to the adhesive matrix. Fillers in general ensure an adequate bond line between the adherends, although due to the porosity of the substrate, low viscosity adhesive fractions tend to penetrate into wood. If penetration occurs, then the Young's modulus of the glue line is increased by the relative increase in filler material content, which can be considered an advantage.

4.2 Assessment of the significance of this study

About 10 years ago the majority view regarding the thermal stability of 1C PUR adhesives was as a disqualifying characteristic for its use in structural applications. Publications by different authors presented results of adhesive tests at elevated temperatures, which revealed deficits in the adhesive strength compared to polycondensation resins, such as PRF or MUF for example. However, investigations within this thesis clearly showed that a one-sided view on this type of adhesive can not sufficiently answer the question if polyurethane adhesives in general are appropriate for use in structural applications at elevated temperatures. The differentiated view on prepolymers, which are the basis for the application oriented adhesives, reveals that as a result of the wide variety of PUR chemistry, a wide variety in the resulting properties is also possible. By systematic modification of the hard segment content and the cross-linking of the prepolymers, the thermal stability could significantly be increased.

In the further formulation of the adhesives it could be shown that the use of selective types of inorganic and organic filler materials also significantly improved the thermal stability of 1C PUR adhesives. These findings were directly incorporated into the ongoing development of new products of the adhesive producers. The new standard for adhesive testing at elevated temperatures in the USA (ASTM D 7247, 2007), which requires a thermal stability of the adhesive that is higher than that of wood, close to its ignition temperature, was in fact a hurdle on which previous polyurethane systems failed. This thesis contributed to the development of a new generation of 1C PUR adhesives that are able to outperform the thermal stability of wood as measured by the aforementioned standard and pass the test.

From the scientific point of view the thesis provides findings in the special field of 1C PUR adhesives for structural applications of engineered wood products that have not been previously published and therefore extends the knowledge of adhesives directly. Furthermore, results of previous publications have been partly reconfirmed, however, some generalized statements regarding the thermal stability of 1C PUR adhesives have been disproved or revised. The close collaboration between the raw material producer Bayer MaterialScience, the adhesive producer Purbond and the wood physics group at ETH has enabled a scientific process that otherwise would hardly be possible.

From these perspectives, the thesis provided a valuable contribution to the development of 1C PUR adhesive for structural applications on the one hand and to the wider usage of the natural recourse of wood on the other.

4.3 Potential for future research

The results of the study have potential to support the development of design models for the fire resistance of bonded structural timber elements, taking into account the behavior of the adhesive at elevated temperatures. Therefore tests on bonded timber elements with optimized adhesive systems are needed. The combination of test results with specimens exposed to fire, as well as pre-heated specimens in an oven, should give a basis to establish a classification and test procedure for structural adhesives with respect to their performance in fire.

4 Synthesis

The current thesis more or less concentrated on the adhesive properties and improvements from the adhesive aspect. Surface machining, as is generally known, also has a significant influence on bonding performance. In future research, interactions between the thermal stability and the method of machining could be an interesting issue. Due to pre-damaged wood cells on the surface, similar interactions are possible than those that have been found in connection with high moisture contents.

By purposive compositions of the prepolymer, it is possible to decrease the urethan hard segments in the matrix and therefore increase the urea hard segment content. This approach turns out to be advantageous at very high temperatures due to the lower thermal stability of the urethane groups. This procedure however results in brittle material behavior and increased foaming during the isocyanate-water reaction. Finding the optimal composition is a potential topic of future research.

The filler material also offers great potential for further research. As described, a multitude of filler types are available that could be tested in combination with 1C PUR. Therefore fillers with functional groups, external fillers or nanoparticle fillers are only a few examples that could improve material properties. In the current thesis, only a rough differentiation of the filler material content was chosen. To find the optimum amount of the tested organic and inorganic fillers used in the project is of great interest for adhesive producers.

The results of the investigations revealed that, in the case of a target-oriented prepolymer configuration and an adjusted formulation, the cohesion of the adhesive system is the predominant problem. Future research on 1C PUR adhesives should therefore concentrate on the adhesion phenomenon that causes reduced bonding strength even at high temperatures. Theories on adhesion are manifold, however the phenomenon of adhesion is still not known in detail.

Several structure-property relationships that are related to the thermal stability of 1C PUR adhesives could be identified, however, their physical and chemical causes partly still remain unanswered up to now. One example is the tempering effect due to which the analyzed prepolymers showed an increase in tensile shear strength from ambient conditions up to temperatures of 70°C. A spectroscopic analysis showed that delayed water reaction can be ruled out for the cause. The DSC and DMA however showed caloric or rheological effects, respectively, that disappear in a second run and therefore suggest a chemical reaction. Morphological investigations on the adhesives could provide further insight into the physical and chemical processes taking place.

4.4 Potential for economical issues

The economic use of the results obtained was an important incitement for the realization of this thesis, which is why the issues of this thesis were economically oriented. The findings were directly integrated into ongoing adhesive developments of the collaborating business companies. As mentioned above, these developments resulted in new products that are able to pass the high standards regarding the thermal stability of the adhesives in North America. Furthermore the results obtained account for the opening up of new markets and the strengthening of the image of 1C PUR adhesives in regions where established adhesives like PRF or MUF are the predominant adhesives used.

Within the framework of this thesis it could be shown where the greatest potential for individual improvements can be found. While the mechanical properties of the adhesive are largely predetermined by the prepolymer composition adaption to the specific wood properties relies on the formulation process. The conclusions obtained help to find starting-points for further developments. The results support the demands of higher cross-linked prepolymers for the adhesive production. The optimal cross-link density of 1C PUR adhesives has to be found in the range between 16 and 20 %, which is above the common standard. The method by which the cross-linking is adjusted is therefore not of importance. This is an important advantage for the adhesive producer, since thereby the possibility exists to adjust the functionality after the prepolymer synthesis.

Different types of filler materials are able to be incorporated within 1C PUR adhesives. This finding helps the adhesive and raw material producers to find economical solutions for the incorporation of the filler material depending on whether a prepolymer-incorporated or an additionally incorporated prepolymer promises an advantage. The first version spares the adhesive producer the complex incorporating process of the filler material, however, the second version provides a high amount of flexibility. Possibly a combined version with an organic prepolymer-incorporated filler with high storage stability and an inorganic additional filler for the final adjustment could be a promising solution.

4.5 Outlook: Some general statements

Present times propose the particular challenge to overcome dependency on finite resources. Wood as an ecological and renewable resource has great potential to significantly contribute towards the success of this ambitious goal. Intelligent research on this evolutionary optimized biocomposite is of great importance for the development of innovative technologies for its material or energetic utilization. Thereby a holistic exploration of their substitution capabilities should be the silver bullet reaching from the forest to wood products and after their product life to a chemical and finally an energetic usage.

Adhesives are an integral part of most different wood-based products, regardless of the degree of digestion or whether timber-based or board material. From this perspective, research on adhesives and even the adhesion of wood is essential to achieve the goals we have set ourselves. 1C PUR adhesives themselves belong to the petrochemical products and, looked at in that light, contradict the requirements. Ongoing research into the substitution of the alcoholic component of 1C PUR by biological sources was intensified in the last years (Pfister, Xia and Larock, 2011). Castor-oil is one example that has been proposed for the use of wood bonding adhesives (Somani et al., 2003).

The ecological aspect is a further important question in this context. Glulam is established as: "an ecologically sound material with favorable disposal properties" (Marutzky, 2002). In general, wood

4 Synthesis

elements such as glulam are made to endure for decades and therefore their disposal is not an ostensible problem. Glulam beams could easily be recycled into particle or fibre boards. For different reasons, the industry refrains from that source and uses predominately round timber. The type of adhesive does not play any role in this regard.

In terms of health compatibility, however, 1C PUR adhesives offer distinct advantages compared to formaldehyde-based adhesives, which account for the dominant adhesive group used in the wood industry. Since formaldehyde was classed as carcinogenic to humans by the IARC (IARC, 2006) in 2004, a lot of effort has been put on decreasing this volatile organic compound (VOC). Since 1C PUR does not contain any solvents or VOCs, glulam manufactured with 1C PUR adhesives is fully comparable with natural solid wood (Marutzky, 2002).

In conclusion, it can be said that increasing usage of the natural renewable recourse of wood, in combination with adhesives that are ecologically and sanitarily unobjectionable, seems to be a step in the right direction, but there is a lot that remains to be done to increase sustainability.

References

Aicher, S. and Reinhardt, H.-W. (2007) Delamination properties and shear strength of glued beech wood laminations with red heartwood. Holz Roh Werkst. 65 No. 2, 125–136

American Forest and Paper Association (2007) Adhesives Awareness Guide. Leesburg, VA: American Wood Council

American Lumber Standard Committee, Inc. (2009) Glued Lumber Policy. Germantown, MD

ASTM D 7247 (2007) Standard test method for evaluating the shear strength of adhesive bonds in laminated wood products at elevated temperatures. West Conshohocken, PA: ASTM International

ASTM D 7374 (2008) Standard practice for evaluating elevated temperature performance of adhesives used in end-jointed lumber. West Conshohocken, PA: ASTM International

ASTM D 7470 (2008) Standard practice for evaluating elevated temperature performance of end-jointed lumber studs. West Conshohocken, PA: ASTM International

ASTM E 119 (2011) Standard test methods for fire tests of building construction and materials. West Conshohocken, PA: ASTM International

Avramidis, G. et al. (2010) Plasma surface modification of wood and wood-based materials. Vak. Forsch. Prax. 22 No. 1, 25–29

Bautechnik (2006) Gutachten zu den Ursachen des Einsturzes der Eissporthalle in Bad Reichenhall vorgelegt. Bautechnik, 83 No. 9, 667–668

Bayer, O. (1947) Das Di-Isocyanat-Polyadditionsverfahren (Polyurethane). Angewandte Chemie, 59 No. 9, 257–272

Bikermann, J.J. (1967) Causes of poor adhesion - Weak boundary layers. Ind. Eng. Chem. 59 No. 9, 40–44

Blanchet, P. (2008) Long-term performance of engineered wood flooring when exposed to temperature and humidity cycling. Forest Prod. J. 58 No. 9, 37–44

Blanchet, P et al. (2003) Comparative study of four adhesives used as binder in engineered wood parquet flooring. Forest Prod. J. 53 No. 1, 89–93

Bramwell, M., editor (1976) The international book of wood. London: Mitchell Beazley, 276 p.

References

Christiansen, A.W. (**2005**) Chemical and mechanical aspects of HMR primer in relationship to wood bonding. Forest Prod. J. 55 No. 11, 73–78

De Bruyne, N.A. and **Houwink, R.**, editors (**1957**) Klebtechnik, Die Adhäsion in Theorie und Praxis. 1st edition. Stuttgart: Berliner Union, 520 p.

DIN 1052 (**2008**) Design of timber structures – General rules and rules for buildings. Berlin: Beuth Verlag

DIN EN 1365-3 (**2000**) Fire resistance tests for loadbearing elements – Part 3: Beams. Berlin: Beuth Verlag

DIN EN 14080 (**2011**) Timber structures – Glued laminated timber and glued solid timber – Requirements. Berlin: Beuth Verlag

DIN EN 15416-2 (**2008**) Adhesives for load bearing timber structures other than phenolic and aminoplastic – Test methods – Part 2: Static load test of multiple bondline specimens in compression shear. Berlin: Beuth Verlag

DIN EN 15416-3 (**2010**) Adhesives for load bearing timber structures other than phenolic and aminoplastic – Test methods – Part 3: Creep deformation test at cyclic climate conditions with specimens loaded in bending shear. Berlin: Beuth Verlag

DIN EN 15425 (**2008**) One component polyurethane for load bearing timber structures – Classification and performance requirements. Berlin: Beuth Verlag

DIN EN 1995-1-1 (**2010**) Eurocode 5: Design of timber structures – Part 1-1: General – Common rules and rules for buildings. Berlin: Beuth Verlag

DIN EN 1995-1-2 (**2010**) Eurocode 5: Design of timber structures – Part 1-2: General – Structural fire design. Berlin: Beuth Verlag

DIN EN 301 (**2006**) Adhesives, phenolic and aminoplastic, for load-bearing timber structures – Classification and performance requirements. Berlin: Beuth Verlag

DIN EN 302-1 (**2004**) Adhesives for load-bearing timber structures – Test methods – Part 1: Determination of bond strength in longitudinal tensile shear strength. Berlin: Beuth Verlag

DIN EN 302-2 (**2004**) Adhesives for load-bearing timber structures – Test methods – Part 2: Determination of resistance to delamination. Berlin: Beuth Verlag

DIN EN 302-3 (**2006**) Adhesives for load-bearing timber structures – Test methods – Part 3: Determination of the effect of acid damage to wood fibres by temperature and humidity cycling on the transverse tensile strength. Berlin: Beuth Verlag

DIN EN 302-4 (**2004**) Adhesives for load bearing timber structures – Test methods – Part 4: Determination of the effects of wood shrinkage on the shear strength. Berlin: Beuth Verlag

Dorn, H. and **Enger, K.** (**1967**) Flame tests on laminated trusses under bending load. Holz Roh Werkst. 25 No. 8, 308–320

Dreyer, R. (1969) Brandverhalten von Holzträgern unter Biege- und Feuerbeanspruchung. Bauen mit Holz, 71 No. 5, 225–227

Fengel, D. (1966) On changes of wood and its components in temperature range up to 200°C – Part II: The hemicelluloses in untreated and thermally treated sprucewood. Holz Roh Werkst. 24 No. 3, 98–109

Follrich, J. et al. (2007a) Tensile strength of softwood butt end joints. Part 2: Improvement of bond strength by a hydroxymethylated resorcinol primer. Wood Mater. Sci. Eng. 2 No. 2, 90–95

Follrich, J. et al. (2007b) Effect of grain angle on shear strength of glued end grain to flat grain joints of defect-free softwood timber. Wood Sci. Technol. 41 No. 6, 501–509

Follrich, J. et al. (2008) Adhesive bond strength of end grain joints in softwood with varying density. Holzforschung, 62 No. 2, 237–242

Fowkes, F. M. (1987) Role of acid-base interfacial bonding in adhesion. J. Adhes. Sci. Technol. 1 No. 1, 7–27

Frangi, A. et al. (2011) Mechanical behaviour of finger joints at elevated temperatures. Wood Sci. Technol., doi:10.1007/s00226-011-0444-9 (in press)

Frangi, A., Fontana, A. and Mischler, A. (2004) Shear behaviour of bond lines in glued laminated timber beams at high temperatures. Wood Sci. Technol. 38 No. 2, 119–126

Furuno, T. et al. (1983) Penetration of glue into the tracheid lumina of softwood and the morphology of fractures by tensile-shear tests. Mokuzai Gakkaishi, 29 No. 1, 43–53

Gavrilovic-Grmusa, I. et al. (2010) Radial penetration of urea-formaldehyde adhesive resins into beech (*Fagus moesiaca*). J. Adhes. Sci. Technol. 24 No. 8-10, 1753–1768

Gerber, S. (2008) Normative Anforderungen an Klebstoffe für den konstruktiven Holzleimbau weltweit. In 2. Kolloquium, Aktuelle Fragen der Holzforschung. Vienna

Gereke, T. et al. (2009) Identification of moisture-induced stresses in cross-laminated wood panels from beech wood (*Fagus sylvatica* L.). Wood Sci. Technol. 43 No. 3-4, 301–315

Gerhards, C.C. (1982) Effect of moisture-content and temperature on the mechanical properties of wood - an analysis of immediate effects. Wood Fiber Sci. 14 No. 1, 4–36

Gindl, W. (2001) SEM and UV-microscopic investigation of glue lines in Parallam PSL. Holz Roh Werkst. 59 No. 3, 211–214

Gindl, W., Dessipri, E. and Wimmer, R. (2002) Using UV-microscopy to study diffusion of melamine-urea-formaldehyderesin in cell walls of spruce wood. Holzforschung, 56 No. 1, 103–107

Gindl, W. et al. (2004) Mechanical properties of spruce wood cell walls by nanoindentation. Appl. Phys. A: Mater. Sci. Process. 79 No. 8, 2069–2073

References

Gindl, W., Schöberl, T. and **Jeronimidis, G. (2004)** The interphase in phenol-formaldehyde and polymeric methylene diphenyl-di-isocyanate glue lines in wood. Int. J. Adhes. Adhes. 24 No. 4, 279–286

Gindl, W. et al. (2005) Direct measurement of strain distribution along a wood bond line. Part 2: Effects of adhesive penetration on strain distribution. Holzforschung, 59 No. 3, 307–310

Glos, P. and **Henrici, D. (1991)** Bending strength and MOE of structural timber (*Picea abies*) at temperatures up to 150°C. Holz Roh Werkst. 49 No. 11, 417–422

Habenicht, G. (2006) Kleben - Grundlagen, Technologien, Anwendungen. 5th edition. Berlin: Springer, 1082 p.

Haß, P. et al. (2009) Influence of growth ring angle, adhesive system and viscosity on the shear strength of adhesive bonds. Wood Mater. Sci. Eng. 4 No. 3, 140–146

Haß, P. et al. (2010) Pore space analysis of beech wood: The vessel network. Holzforschung, 64 No. 5, 639–644

Haß, P. et al. (2012) Adhesive penetration in beech wood: Experiments. Wood Sci. Technol. 46 No. 1-3, 243–256

Horvath, N., Molnar, S. and **Niemz, P. (2007)** Untersuchungen zum Einfluss der Holzfeuchte auf ausgewählte Eigenschaften von Fichte, Eiche und Rotbuche. Holztechnologie, 49 No. 1, 10–15

IARC (2006) Formaldehyde, 2-Butoxyethanol and 1-tert-Butoxypropan-2-ol. Volume 88, IARC Monographs on the Evaluation of Carcinogenic Risks to Humans. International Agency for Research on Cancer, International Agency for Research on Cancer, 16 p.

Jakes, J. E. et al. (2008) Experimental method to account for structural compliance in nanoindentation measurements. J. Mater. Res. 23 No. 4, 1113–1127

Johnson, S. E. and **Kamke, F. A. (1992)** Quantitative-analysis of gross adhesive penetration in wood using fluorescence microscopy. J. Adhes. 40 No. 1, 47–61

Kägi, A., Niemz, P. and **Mandallaz, D. (2006)** Influence of moisture content and selected technological parameters on the adhesion of one-part polyurethane adhesives under extremeclimatical conditions. Holz Roh Werkst. 64 No. 4, 261–268

Källander, B. and **Lind, P. (2006)** Strength properties of wood adhesives after exposure to fire. In **Frihart, C. R.**, editor: Wood Aadhesives 2005. Madison, WI: Forest Products Society, 211–219

Kamke, F. A. and **Lee, J. N. (2007)** Adhesive penetration in wood - a review. Wood Fiber Sci. 39 No. 2, 205–220

Kemmsies, M. (1999) Comparison of bond lines in glulam beams adhered with a phenol-resorcinol-formaldehyde (PRF) and one-component polyurethane (PUR) after fire exposure. In **Boström, L.**, editor: First RILEM Symposium on Timber Engineering. Volume 8, Bagneux: RILEM Publications SARL, 399–408

Kolb, H. (1968) Tests on bending strength and fire tests with glulam-girders of beech veneers. Holz Roh Werkst. 26 No. 8, 277–283

Kollmann, F. (1951) Technologie des Holzes und der Holzwerkstoffe. 2nd edition. Berlin: Springer, 2233 p.

König, J. (2005) Structural fire design according to Eurocode 5 - design rules and their background. Fire Mater. 29 No. 3, 147–163

König, J., Norén, J. and Sterley, M. (2008) Effect of adhesives on finger joint performance in fire. In Working commission W18 - Timber Structures. International Council for Research and Innovation in Building and Construction (CIB)

Konnerth, J. and Gindl, W. (2006) Mechanical characterisation of wood-adhesive interphase cell walls by nanoindentation. Holzforschung, 60 No. 4, 429–433

Konnerth, J. et al. (2008) Adhesive penetration of wood cell walls investigated by scanningthermal microscopy (SThM). Holzforschung, 62 No. 1, 91–98

Konnerth, J. et al. (2010) Elastic properties of adhesive polymers. III. Adhesive polymer films under dry and wet conditions characterized by means of nanoindentation. J. Appl. Polym. Sci. 118 No. 3, 1331–1334

Konnerth, J., Valla, A. and Gindl, W. (2007) Nanoindentation mapping of a wood-adhesive bond. Appl. Phys. A: Mater. Sci. Process. 88 No. 2, 371–375

Korley, L.T.J. et al. (2006) Effect of the degree of soft and hard segment ordering on the morphology and mechanical behavior of semicrystalline segmented polyurethanes. Polymer, 47 No. 9, 3073–3082

Kudela, J. (1996) Influence of moisture and temperature loading on strength of beech wood loaded in compression. Drev. Vysk. 41 No. 2, 3–17

Marutzky, R. (2002) Brettschichtholz – Hochwertiger und ökologisch unbedenklicher Baustoff mit günstigen Entsorgungseigenschaften. Bauen mit Holz, 104 No. 2, 30–33

Modzel, G., Kamke, F. and De Carlo, F. (2011) Comparative analysis of a wood: adhesive bondline. Wood Sci. Technol. 45 No. 1, 147–158

MPA-Liste (2011) Klebstoffliste I der MPA Universität Stuttgart betreffend geprüfter Klebstoffe im Geltungsbereich der DIN 1052 und allgemeiner bauaufsichtlicher Zulassungen. Stuttgart

Müller, R. (2000) Verhalten von auf Schub beanspruchten BSH-Leimfugen bei hohen Temperaturen. Master's thesis, ETH Zurich, IBK

Na, B. et al. (2005) One-component polyurethane adhesives for green wood gluing: Structure and temperature-dependent creep. J. Appl. Polym. Sci. 96 No. 4, 1231–1243

Niemz, P. (1993) Physik des Holzes und der Holzwerkstoffe. Leinfelden-Echterdingen: DRW-Verlag, 243 p.

References

Niemz, P. (2006) Einfluss der Temperatureinwirkung auf Klebstoffe. Holz-Zentralblatt, No. 30, 878

Niemz, P. and **Allenspach, K. (2009)** Studies on the influence of temperature and timber moisture on the failure behaviour of selected adhesives under tensile shear load. Bauphysik, 31 No. 5, 296–304

Niemz, P. et al. (2004) Experiments on the distribution of adhesives close to the glue joint by neutron radiography and microscopy. Holz Roh Werkst. 62 No. 6, 424–432

Nussbaum, R.M. and **Sterley, M. (2002)** The effect of wood extractive content on glue adhesion and surface wettability of wood. Wood Fiber Sci. 34 No. 1, 57–71

Östman, B.A.-L. (1985) Wood tensile strength at temperatures and moisture contents simulating fire conditions. Wood Sci. Technol. 62 No. 6, 424–432

Pfister, D.P., Xia, Y. and **Larock, R.C. (2011)** Recent advances in vegetable oil-based polyurethanes. ChemSusChem, 4 No. 6, 703–717

Pizzi, A., Mtsweni, B. and **Parsons, W. (1994)** Wood-induced catalytic activation of PF adhesives autopolymerization vs. PF/wood covalent bonding. J. Appl. Polym. Sci. 52 No. 13, 1847–1856

Pizzi, A. and **Owens, N.A. (1995)** Interface covalent bonding vs wood-induced catalytic autocondensation of diisocyanate wood adhesives. Holzforschung, 49 No. 3, 269–272

Pommier, R. and **Elbez, G. (2006)** Finger-jointing green softwood: Evaluation of the interaction between polyurethane adhesive and wood. Wood Mater. Sci. Eng. 1 No. 3, 127–137

Rapp, A.O. et al. (1999) Electron energy loss spectroscopy (EELS) for quantification of cell-wall penetration of a melamine resin. Holzforschung, 53 No. 2, 111–117

Richter, K., Pizzi, A. and **Despres, A. (2006)** Thermal stability of structural one-component polyurethane adhesives for wood - Structure-property relationship. J. Appl. Polym. Sci. 102 No. 1, 24–32

Richter, K. and **Schirle, M. (2002)** Behaviour of 1K PUR adhesives under increased moisture and temperature conditions. Lignovisionen, No. 4, 149–154

Richter, K. and **Steiger, R. (2005)** Thermal stability of wood-wood and wood-FRP bonding with polyurethane and epoxy adhesives. Adv. Eng. Mater. 7 No. 5, 419–426

Rijckaert, V., Stevens, M. and **Acker, J. van (2001)** Effect of some formulation parameters on the penetration and adhesion of water-borne primers into wood. Holz Roh Werkst. 59 No. 5, 344–350

Rijckaert, V et al. (2001) Quantitative assessment of the penetration of water-borne and solvent-borne wood coatings in Scots pine sapwood. Holz Roh Werkst. 59 No. 4, 278–287

Robertson, R. E. (1975) Strength of an adhesive weak boundary-layer. J. Adhes. 7 No. 2, 121–136

Schirle, M. et al. (2002) Charakterisierung und Optimierung der Holzverklebung mit 1 Komponenten Polyurethan (1K-PUR) Klebstoffen. Dübbendorf: EMPA Abteilung Holz (4126.1). – KTI Abschlussbericht

Schmidt, M. (2001) Verhalten von auf Schub beanspruchten BSH-Leimfugen bei hohen Temperaturen. Master's thesis, ETH Zurich, IBK

Schmidt, M., Glos, P. and Wegener, G. (2010) Gluing of European beech wood for load bearing timber structures. Eur. J. Wood Prod. 68 No. 1, 43–57

Schneider, A. (1971) Investigations on the influence of heat treatments within a range of temperature from 100 to 200°C on the modulus of elasticity, maximum crushing strength, and impact work of pine sapwood and beech wood. Holz Roh Werkst. 29 No. 11, 431–440

Schrödter, A. and Niemz, P. (2006) Investigation on the failure behaviour of glue joints at high temperatures and relative humidity. Holztechnologie, 47 No. 1, 24–32

Šebenik, U. and Krajnc, M. (2007) Influence of the soft segment length and content on the synthesis and properties of isocyanate-terminated urethane prepolymers. Int. J. Adhes. Adhes. 27 No. 7, 527–535

Sernek, M., Resnik, J. and Kamke, F.A. (1999) Penetration of liquid urea-formaldehyde adhesive into beech wood. Wood Fiber Sci. 31 No. 1, 41–48

Sharpe, L.H. and Schonhorn, H.; Fowkes, F.M., editor (1964) Chap. Surface energetics, adhesion and adhesive joints In Contact angle, wettability and adhesion. Volume 43, Washington, DC: American Chemical Society, 189–201

Somani, K.P. et al. (2003) Castor oil based polyurethane adhesives for wood-to-wood bonding. Int. J. Adhes. Adhes. 23 No. 4, 269–275

St-Pierre, B. et al. (2005) Effect of moisture content and temperature on tension strength of fingerjointed black spruce lumber. Forest Prod. J. 55 No. 12, 9–16

Stehr, M., Gardner, D.J. and Wålinder, M.E.P. (2001) Dynamic wettability of different machined wood surfaces. J. Adhes. 76 No. 3, 185–200

Stehr, M. and Johansson, I. (2000) Weak boundary layers on wood surfaces. J. Adhes. Sci. Technol. 14 No. 10, 1211–1224

Sterley, M., Bümer, H. and Wålinder, M. E. P. (2004) Edge and face gluing of green timber using a one-component polyurethane adhesive. Holz Roh Werkst. 62 No. 6, 479–482

Sterley, M. and Gustafson, P.J. (2006) Shear fracture properties of green-glued polyurethane wood adhesive bonds. In Frihart, C.R., editor: Wood Adhesives 2005. Madison, WI: Forest Products Society, 221–229

Suchsland, O. (1958) On the penetration of glue in wood gluing and the significance of the penetration depth for the strength of glue joints. Holz Roh Werkst. 16 No. 3, 101–108

References

Świetliczny, M. (1980) Über den Einfluß der Neigung der Holzfasern auf die Festigkeit der Klebfugen. Holztechnologie, 21 No. 2, 83–87

Szalai, J. et al. (2004) Untersuchungen zum Einfluss der Holzfeuchtigkeit auf das Bruchverhalten von Fichte bei Zugbelastung in Faserrichtung. Schweiz. Z. Forstwes. 155 No. 1, 1–5

Uhlig, K. (2006) Polyurethan Taschenbuch. 3rd edition. München: Hanser, 208 p.

Vick, C.B. and Okkonen, E.A. (1998) Strength and durability of one-part polyurethane adhesive bonds to wood. Forest Prod. J. 48 No. 11-12, 71–76

Vick, C.B. and Okkonen, E.A. (2000) Durability of one-part polyurethane bonds to wood improved by HMR coupling agent. Forest Prod. J. 50 No. 10, 69–75

Voyutskii, S.S. (1963) Autohesion and adhesion of high polymers. New York: Interscience Publishers, Polymer reviews, 272 p.

Wålinder, M.E.P. and Johansson, I. (2001) Measurement of wood wettability by the Wilhelmy method - Part 1. Contamination of probe liquids by extractives. Holzforschung, 55 No. 1, 21–32

Wallström, L. and Lindberg, K.A.H. (1999) Measurement of cell wall penetration in wood of water-based chemicals using SEM/EDS and STEM/EDS technique. Wood Sci. Technol. 33 No. 2, 111–122

Wang, W.Q. and Yan, N. (2005) Characterizing liquid resin penetration in wood using a mercury intrusion porosimeter. Wood Fiber Sci. 37 No. 3, 505–513

White, M.S. (1977) Influence of resin penetration on fracture toughness of wood adhesive bonds. Wood Science, 10 No. 1, 6–14

White, M.S., Ifju, G. and Johnson, J.A. (1977) Method for measuring resin penetration into wood. Forest Prod. J. 27 No. 7, 52–54

White, R.H. and Dietenberger, M.A.; Buschow, K.H.J. et al., editors (2001) Chap. Wood Products: Thermal degradation and fire. In Encyclopedia of materials: Science and technology. 2nd edition. Oxford: Elsevier Science Ltd., 9712–9716

Widsten, P. et al. (2006) Factors influencing timber gluability with one-part polyurethanes, studied with nine Australian timber species. Holzforschung, 60 No. 4, 423–428

Wittmann, O. (1976) Gluing wood with isocyanate. Holz Roh Werkst. 34 No. 11, 427–431

Wurtz, C.A. (1848) Recherches sur les éthers cyaniques et sur le cyanurate de méthylene. C. R. Hebd. Seances Acad. Sci. 27, 241–243

Yang, T.H. et al. (2009) Effect of fire exposure on the mechanical properties of glued laminated timber. Mater. Design, 30 No. 3, 698–703

Yeh, B. et al. (2006) Adhesive performance at elevated temperatures for engineered wood products. In **Friehard, C.R., editor:** Wood Adhesives 2005. Madison, WI: Forest Products Society, 195–201

Zeppenfeld, G. and **Grunwald, D.** (**2005**) Klebstoffe in der Holz- und Möbelindustrie. 2nd edition. Leinfelden-Echterdingen: DRW-Verlag, 368 p.

i want morebooks!

Buy your books fast and straightforward online - at one of world's fastest growing online book stores! Environmentally sound due to Print-on-Demand technologies.

Buy your books online at
www.get-morebooks.com

Kaufen Sie Ihre Bücher schnell und unkompliziert online – auf einer der am schnellsten wachsenden Buchhandelsplattformen weltweit! Dank Print-On-Demand umwelt- und ressourcenschonend produziert.

Bücher schneller online kaufen
www.morebooks.de

VDM Verlagsservicegesellschaft mbH
Heinrich-Böcking-Str. 6-8   Telefon: +49 681 3720 174   info@vdm-vsg.de
D - 66121 Saarbrücken   Telefax: +49 681 3720 1749   www.vdm-vsg.de

Printed by Books on Demand GmbH, Norderstedt / Germany